What Every Engineer Should Know about Python

Engineers across all disciplines can benefit from learning Python. This powerful programming language enables engineers to enhance their skill sets and perform more sophisticated work in less time, whether in engineering analysis, system design and development, integration and testing, machine learning and other artificial intelligence applications, project management, or other areas. *What Every Engineer Should Know About Python* offers students and practicing engineers a straightforward and practical introduction to Python for technical programming and broader uses to enhance productivity. It focuses on the core features of Python most relevant to engineering tasks, avoids computer science jargon, and emphasizes writing useful software while effectively leveraging generative AI.

- Features examples tied to real-world engineering scenarios that are easily adapted
- Explains how to leverage the vast ecosystem of open-source Python packages for scientific applications, rather than developing new software from scratch
- Covers the incorporation of Python into engineering designs and systems, whether web-based, desktop, or embedded
- Provides guidance on optimizing generative AI with Python, including case study examples
- Describes software tool environments and development practices for the rapid creation of high-quality software
- Demonstrates how Python can improve personal and organizational productivity through workflow automation
- Directs readers to further resources for exploring advanced Python features

This practical and concise book serves as a self-contained introduction for engineers and readers from scientific disciplines who are new to programming or to Python.

Raymond J. Madachy, Ph.D., is a Professor in the Systems Engineering Department at the Naval Postgraduate School. He has worked in academia, industry, and consulting, serving in both technical and managerial roles. He earned his B.S. in Mechanical Engineering from the University of Dayton, his M.S. in System Science from the University of California, San Diego, and his Ph.D. in Industrial and Systems Engineering from the University of Southern California.

His research has been funded by diverse agencies across the Department of Defense, the National Security Agency, NASA, and various companies. His research interests include systems engineering tool environments for digital engineering, modeling and simulation of systems and software engineering processes, including generative AI usage, integrating systems and software engineering disciplines, system cost modeling, and affordability and tradespace analysis. He has developed widely used cost estimation tools for systems and software engineering, and serves as the lead developer of the open-source Systems Engineering Library (*se-lib*).

He previously authored *Software Process Dynamics* and *What Every Engineer Should Know about Modeling and Simulation,* and co-authored *Software Cost Estimation with COCOMO II* and *Software Cost Estimation Metrics Manual for Defense Systems.*

What Every Engineer Should Know

Series Editor

Phillip A. Laplante
Pennsylvania State University

For more information about this series, please visit:
www.routledge.com/What-Every-Engineer-Should-Know/book-series/CRCWEESK

What Every Engineer Should Know About Python

Raymond J. Madachy

CRC Press
Taylor & Francis Group
Boca Raton London New York

CRC Press is an imprint of the
Taylor & Francis Group, an **informa** business

First edition published 2025
by CRC Press
2385 NW Executive Center Drive, Suite 320, Boca Raton FL 33431

and by CRC Press
4 Park Square, Milton Park, Abingdon, Oxon, OX14 4RN

CRC Press is an imprint of Taylor & Francis Group, LLC

Library of Congress Cataloging-in-Publication Data
Names: Madachy, Raymond J. (Raymond Joseph), author.
Title: What every engineer should know about Python / Raymond J. Madachy.
Identifiers: LCCN 2024050063 (print) | LCCN 2024050064 (ebook) | ISBN
9781032355627 (hbk) | ISBN 9781032358185 (pbk) | ISBN 9781003331070
(ebk)
Subjects: LCSH: Computer-aided engineering. | Python (Computer program
language) | Computer programming.
Classification: LCC TA345 .M327 2025 (print) | LCC TA345 (ebook) | DDC
005.13/3--dc23/eng/20250123
LC record available at https://lccn.loc.gov/2024050063
LC ebook record available at https://lccn.loc.gov/2024050064

ISBN: 978-1-032-35562-7 (hbk)
ISBN: 978-1-032-35818-5 (pbk)
ISBN: 978-1-003-33107-0 (ebk)

DOI: 10.1201/9781003331070

Typeset in Nimbus Roman font
by KnowledgeWorks Global Ltd.

Publisher's Note
This book has been prepared from camera-ready copy provided by the author.

Dedication

This book wouldn't exist without the many past pioneers of computing.

Contents

List of Figures

List of Tables

List of Listings

Foreword

As an engineer with over 25 years in industry and 15 years in academia, spanning a wide range of domains from supercomputing to complex enterprise systems, I've had a front-row seat to the evolution of technology and the growing importance of data and computation in driving informed, adaptive engineering practices. Entering Industry 5.0, we see data as a powerful catalyst that accelerates the OODA loop (observe, orient, decide, and act) in real-time—reshaping not only how we understand the world but also how we design, troubleshoot, and innovate.

Python has emerged as the common language of this data-driven revolution, providing engineers with a flexible, high-level programming environment that is both intuitive and immensely powerful. It is fair to say Python has become the "Esperanto" of data sciences, AI, and machine learning, empowering professionals across disciplines to speak a common language, share solutions, and drive advancements in ways previously unimaginable. For engineers, especially, Python is no longer just a coding skill but a central tool in our arsenal.

What Every Engineer Should Know About Python is a timely and invaluable resource, precisely because it balances the essential principles of Python with its real-world application across various engineering practices. It equips engineers with not only the syntax and structure of Python but also the robust utility libraries, scientific tools, and engineering-specific applications that can accelerate and enhance our work.

This book serves as a complete reference for engineers to find everything in one place with explicit examples. It begins with the basics, guiding readers from foundational Python concepts to more advanced programming techniques, including the use of powerful libraries like NumPy, Matplotlib, and Pandas. These chapters ensure that every engineer, regardless of their familiarity with Python, can quickly become competent and confident in using it. Further, the book covers complex topics like object-oriented programming, embedded systems, and even cutting-edge applications in machine learning, robotics, and web-based interfaces.

In addition to providing hands-on examples like projectile motion and wind turbine power simulation, this book also takes a structured dive into Python applications for critical engineering analysis—from statistical studies to simulations of robotic paths and manufacturing processes. Each chapter is carefully designed to build practical skills that bridge theoretical concepts with their real-world engineering counterparts.

As we move towards a future where young engineers will increasingly work on transdisciplinary projects, this book also introduces modeling notations from systems and software engineering to facilitate cross-discipline communication. This forward-looking approach will undoubtedly help engineers adapt to evolving industry demands.

I wholeheartedly endorse this book as an important contribution to engineering education and look forward to my students using it as a reference. Ultimately, *What Every Engineer Should Know About Python* does more than teach Python syntax; it equips engineers to harness Python's full potential across diverse fields.

Ray Madachy has provided a resource that will empower engineers to use Python effectively, not only to solve problems but also to design innovative systems for the future. I am excited to see how this book will empower the next generation of engineers to not only accelerate the OODA loop but to redefine the very boundaries of what is possible in engineering today.

Dr. Jon Wade

Professor of Practice, Director of Convergent Systems Engineering, Jacobs School of Engineering, University of California, San Diego; Fellow, International Council of Systems Engineering (INCOSE)

Preface

Python is a powerful high-level general-purpose programming language that can be applied to many types of problems. It uses a clear, elegant syntax, making programs easy to write, read, and maintain. It has many advanced features and supports both object-oriented and functional programming paradigms. It has become a dominant language in engineering due to its versatility, ease of use, and powerful ecosystem of open-source libraries.

But in these days of generative AI tools that help write software, does an engineer still need to understand the fundamentals of a programming language and its environment? The answer is a resounding YES. AI assistants cannot yet be fully trusted to produce correct programs for several reasons. They can be useful when prompted with specific language terminology, but their outputs must be carefully evaluated, thoroughly tested, modified as necessary, and re-tested. This requires knowing how to use the necessary tools in the development environment.

To successfully leverage AI assistance, one must understand what is feasible with Python and how to express it. This book provides an overview of what can be achieved in a variety of engineering applications and scenarios. It was initially started before generative AI became widely available, but guidance on how to effectively use AI assistants with Python has been added.

Python's intuitive syntax allows engineers to focus on solving problems and applying solution principles with fewer distractions. Its expressiveness expands an engineer's capabilities, allowing them to work at a higher level of abstraction. The language uses standard English keywords, which makes it easier to write code that more closely resembles the concise mathematics used to solve engineering problems.

The reader will learn Python with the immediate goal of writing useful programs for engineering purposes. The book demonstrates professional-level language features and data structures, and it is self-contained for readers who have never programmed before. Pointers to more extensive material online are also provided, so readers will know where to look for further information and how to evaluate AI-generated content.

Python is free to download, use, and include in applications. It is portable across platforms and operating systems, and can run on almost any environment (desktop, cloud, devices, etc.). It is compatible with macOS, Windows, Linux, other Unix variants, iOS, Android, Raspberry Pi OS, and many others. Python also includes a large standard library that supports common tasks such as operating system calls, window systems, web interactions, automated testing, and more.

In addition to its traditional use in numerical analysis (a focus of other Python books for engineers and scientists), Python can be integrated into systems, product designs, and workflows to improve personal productivity in various ways.

Python provides functionality for many types of systems and embedded devices, from on-board flight software to Internet of Things (IoT) applications, dynamic websites, and phone apps. Knowing Python's capabilities equips engineers with insight into design options and trade-offs.

As an example of workflow automation, this book was developed using Python scripts to generate LaTeX markup and automatically recreate all the included graphics files when code or global styles were updated. Given the many change iterations, numerous links throughout the text, and the file management and regression testing required for all examples, this automation was not overkill.

The author first learned Fortran as an undergraduate and has used C, Basic, PHP, Java, and other languages. What might take days to implement in another language can often be reduced to hours using Python, with far fewer lines of code. The benefits are even greater when considering the ease of maintaining Python code, thanks to its emphasis on readability.

The bottom line is that engineers can harness the power of Python's extensive libraries to become more efficient, perform more sophisticated analyses, work more rapidly, integrate systems more effectively, and develop better solutions for the challenges confronting humanity.

AUDIENCE

This book is for engineering or science students new to programming, working engineers without recent programming experience, and those currently proficient with programming (who may be motivated to use a better language). It is a self-contained introduction and does not assume any programming experience, though any previous exposure is useful. Readers already up the learning curve can skip ahead past the basics.

Students will be well positioned to apply Python in future courses, academic projects, and their careers. This will be a distinct advantage. If the reader is a working professional, then they can update and improve their skillset for greater challenges and further advancement.

BOOK CONTENTS

Chapter 0 introduces the underlying philosophy and principles of Python, then starts the reader with an introduction to environments and basic syntax. It explains how Python works as an interpreted language within tool environments including command line consoles, editors, and comprehensive development environments. Open source tools are covered for code and documentation development in the Jupyter Notebook environment and other editors.

Initial language syntax and features are defined to get up and running immediately. An initial example for a recurring trajectory analysis problem is introduced to demonstrate what it means to be "Pythonic". The reader can understand the example before a formal introduction to the full language syntax as a testament to the legibility of Python.

Chapter 1 introduces the language syntax with ample illustrations from engineering scenarios. It covers the definitions of variables, data types, and structures.

Language elements are described for logical operators, conditions and if statements, iteration (for which there are many iterable types of structures), other handy functions, printing, and working with files. These elements are building blocks to compose larger programs with.

Functions are introduced which are fundamental building blocks. They serve as reusable code pieces that streamline programming and improve overall quality. The chapter describes how to create functions and specify the arguments that are passed between functions, which may include other functions (composition of functions).

Python implements object-oriented programming using classes. A concise overview of object oriented concepts is provided, then the implementation of classes, their attributes and methods with examples. These are contrasted to functions as another way to solve the same problems. Using classes isn't necessary, but the practice is pervasive, and much open source software is written and used this way. It behooves one to understand Python classes for this reason alone.

Chapter 2 introduces the primary Python libraries that are important for scientific applications. They are general purpose and usable in any field of engineering. The major ones covered are NumPy for basic numerical computation (which is used by many other libraries), SciPy for scientific applications, Matplotlib for visualization, and Pandas for working with datasets. NetworkX is covered for applications of discrete network and tree structures, and Graphviz for graph visualization of structures.

Engineering examples illustrate the primary features and operations of each library. These examples build upon one another, allowing for the integration of capabilities from multiple libraries in later sections." These core libraries are then used in subsequent examples in Chapters 3 and 4.

Chapter 3 showcases examples for broad engineering analysis applications building on previous chapters. They are traditional applications in major disciplines. They demonstrate integration of language features, core libraries, and some application specific libraries. First are general methods for statistical analysis followed by examples in a variety of fields.

Some problems are solved analytically and others require simulation. They include both static and dynamic problems, continuous and discrete computation. Modeling and simulation are covered including Monte Carlo analysis. The examples are easily adapted and extended for other contexts.

Chapter 4 moves away from language features to building systems with Python. It demonstrates straightforward examples for incorporating Python into engineered systems. Python can be embedded in physical devices (e.g., web-enabled Internet of Things), programmed to control drones and autonomous devices, serve as a web application, run web servers, manage applications locally or across a network, and integrate data across a large enterprise.

There are other ways to interact with Python besides an IDE. Applications can be made friendlier for engineers and other stakeholders. Applications can run in a browser, desktop or mobile device. As shown, web applications will require specific libraries or inter-operation with other languages. Tools and libraries to incorporate graphical user interfaces for Python desktop programs are demonstrated.

Python may operate over a network or in a device. It is capable of many network operations, both locally and on the internet. Examples are explained for generating static web sites and executable web applications. Others are shown that execute on embedded systems to control hardware device operations. The Raspberry Pi platform is demonstrated, but there are many others that work similarly.

These capabilities are demonstrated for small-scale engineering applications but can be readily scaled up. Core templates and generalized routines are provided as foundations to create more complex applications.

Chapter 5 focuses on processes and tools essential for effective Python software development, with a primary emphasis for individuals and introducing best practices that become increasingly important in team environments. Adopting best practices and leveraging appropriate tools in a disciplined engineering approach will enhance productivity, improve software quality, and support collaboration on multidisciplinary teams.

Examples are shown of how Python can be integrated with command line operations to automate essential tasks. Automating such processes not only saves time but also reduces the likelihood of human error. By combining command line tools with Python scripts, engineers can create more efficient workflows that streamline repetitive work, allowing them to focus more on complex aspects.

The chapter also covers iterative development explaining how to evolve a program in short cycles, automated testing, debugging, and documentation. Process guidance on using generative AI is provided and demonstrated with an extended case study example.

Tools include IDEs for development and platforms for code management. Methods for debugging and testing on individual IDEs are shown. Important considerations for performance, speed, and memory tradeoffs are described. Guidance is provided for different contexts depending on system goals.

Team processes are overviewed for collaborative configuration management, version control, continuous integration, and deployment that are critical when working on large projects. Using GitHub or similar platforms is explained regarding how they support individual and collaborative team work.

Engineering workflows can also be automated with Python and tools as demonstrated. It can streamline activities for documentation and analysis, integration of systems and data, or other tasks. It can support personal workflow automation up to enterprise-wide integration. Continuous tool improvement is emphasized to keep up with technology and to automate more engineering tasks.

Finally, Appendices are provided for reference with supplemental language details. These augment some of the lists and descriptions provided in Chapter 1 for a more complete reference.

ACKNOWLEDGMENTS

This book would not have been possible without the support and assistance of many people. Many thanks for the peer reviews and helpful ideas from Dr. Barclay Brown at Rockwell Collins, Dr. Timothy Ferris at Cranfield University, Dr. Mike Green at

Naval Postgraduate School (NPS), PhD students Ryan Bell and Ryan Longshore at NPS, and Jay Mackey. Dr. Ron Giachetti initially encouraged me to use Python in our NPS curriculum, and Dr. Oleg Yakimenko at NPS has been very supportive. The most enduring conversations were had with Joe Raby at IBM who was a captive audience while hiking hundreds of miles in the mountains.

Allison Shatkin at CRC Press has been instrumental throughout the writing process. She was always responsive to help while enduring mid-course volatility and more than a few delays. I'm indebted for her great patience and unflagging support.

My wife Nancy endured the writing of yet another book with lots of patience. She supported me and sacrificed our time together to let me finish it. She was critical, and I am most grateful.

FOR INSTRUCTORS

This book is primarily designed to serve as a text for undergraduate engineering students and professionals new to Python. To ease the learning curve, the examples are small and relatively simple applications. The exercises range from basic to more complex problems. Some advanced exercises may also be suitable for graduate-level work, and the examples can be extended accordingly.

Different paths through the topics are possible for various course lengths and levels. A course may focus on engineering analysis, as emphasized in Chapter 3, and/or the engineering of systems as presented in Chapter 4. Instead of diving deep into specific engineering areas, the book showcases a multitude of functionalities and platforms available with Python. Pointers and references are provided for additional libraries and types of applications not covered in the chapters.

The book is designed for students in any engineering discipline and takes the perspective that they will be working with other disciplines as professionals. However, if students are in software engineering, it is best suited for an initial course before they learn more about processes and other languages. In the spirit of transdisciplinary engineering, modeling methods from system and software engineering are introduced to (1) better explain some Python operations and (2) help bridge communication and analyses on real projects.

Instructors may access supplemental course materials to facilitate teaching at CRC Press's book website. Jupyter notebooks and standalone programs are available for course usage. They cover the book chapters with slides, exercise solutions, and extended examples that did not make the book due to size restrictions. The published content is only a fraction of the full set of examples. Errata and other new materials will be made available.

Instructors will also have access to book updates referred to in the next section. They are encouraged to influence the updates by giving feedback to the author based on their teaching experiences.

BOOK UPDATES

Needless to say, technology is advancing faster all the time. Not all recent developments could make it in this book in order to finish it. However, some emerging areas are evident for a next edition. These include using Python with Large Language Models (LLMs) for generative AI (which the author is already undertaking), more of a focus on modern web frameworks, and mobile device applications. It is an exciting time that calls for continuous learning.

Comments on this book and experiences with Python are of great interest to the author, and feedback will help in developing the next edition. You are encouraged to send any ideas, improvement suggestions, new and enhanced examples, or exercises. They may be incorporated in future editions with due credit given.

0 Introduction

A Python program, also called a script, is a list of statements typically contained in a file or in a computation cell. A *statement* is an instruction that a Python interpreter can execute. The interpreter is a program normally called "python" that will run Python programs on a computer.

When the statements are executed, they are read sequentially and processed accordingly by the interpreter. Statements may contain *expressions* which are combinations of operands (e.g., constants and variables) and operators (e.g., +, -, *, /) that produce values. The interpreter evaluates the expressions to compute values that may be numeric, logical flags, character strings, sequences, or otherwise.

A program typically does the following in order to solve engineering problems:

- Read input from the keyboard or files

- Perform computations

- Make decisions based on logical conditions

- Repeat the same operations for different inputs

- Write output to the screen or files

0.1 HELLO WORLD EXAMPLE

A *hello world* example for a language demonstrates the minimum steps to write and execute a program that prints "hello world" or similar output. It is usually a first student exercise to test the environment. Below is our single line example in the gray code block with syntax coloring[1]. When it is run, the printed output is displayed as shown underneath the code.

```python
print("Hello engineers around the world!")
```
```
Hello engineers around the world!
```

In the code statement, `print` is a special built-in Python function to display the value of the expression provided in the parentheses. The expression is the *argument* passed to the function for its processing which is a character string enclosed in double quotes.

The print statement will run directly in a command line interpreter after hitting the enter key, or it can be run as a separate program from an editor that sends it to the interpreter for execution. Examples in different environments are shown next.

---

[1]The syntax coloring used in this book is the Spyder editor light color scheme which shows built-in functions in magenta and character strings in green.

0.2 GETTING STARTED

There are many Python interpreters and Integrated Development Environments (IDEs) available, both desktop and cloud-based. One can choose different environments based on specific needs and even use them in combination, as they often offer complementary features. All environments provide an interpreter to run Python commands directly in a command line (also called a console interface). IDEs have code editors and are able to execute entire programs. Table 0.1 summarizes differences between desktop and cloud-based IDE platforms, where a platform is a combination of software and hardware.

An IDE allows one to both write and run Python scripts within the same interface. It typically combines a code editor, an interactive Python console, and additional window panes, such as for help. The code editor enables one to write and edit scripts with features like syntax highlighting, code completion, and debugging. The interactive Python console acts as an interpreter, allowing commands to be executed one at a time with immediate results. This console usually replaces the need for a separate terminal window for most tasks.

Open source tools are widely used in the Python community. Modern IDEs are built with the open-source IPython console, an enhanced console with advanced features for syntax highlighting, more informative error messages, inline plotting, and shell command access. Many IDEs also include features such as library lookups, AI assistance, GitHub integration, auto-documentation, and customization options. These components work together to provide a unified environment.

Table 0.1 includes *deployment diagrams* of local and cloud-based platforms, illustrating how software components are physically deployed on hardware. All resources are provided for on a desktop, and for cloud-based, a browser is required to access resources on web servers communicating over the Internet using the http protocol. The diagrams from the Unified Modeling Language (UML) visualize how the IDE Python interpreters and working code files are distributed across different physical devices, and their connections. Nodes for physical devices and software execution environments appear as boxes, where different node types are indicated as stereotypes surrounded by guillemets (<< >>). An interpreter serves as an execution environment because it runs the Python code residing on the same device (see the next section). Devices contain the physical computing resources with processing memory and services necessary to run the interpreter.

Any of the environments listed below is sufficient to get started, though the level of sophistication in the code editors and available tools will vary. A bare minimum setup can be achieved with any text editor and basic Python installation. In some cases such as embedded hardware, this might be the only choice of development platform with limited or no debugging support.

0.2.1 INTERPRETER OVERVIEW

A Python interpreter is software that reads and executes commands interactively. It provides the runtime environment for code to execute, as well as providing access to

Table 0.1

Comparison of Python Desktop vs. Cloud-Based IDE Platforms

Aspect	Desktop	Cloud-Based
Requirements	Python, IDE	Web browser, internet connection, cloud account
Deployment		
File management	Permanent storage of all files (code, input data, output files) on local machine.	Files being developed are saved in the cloud. Files uploaded or generated during a session are ephemeral unless saved explicitly.
Collaboration	None or limited. Requires version control systems or manual sharing of files.	Built-in real-time collaboration features.
Customization	Highly customizable	Limited customization with restrictions on package installation and environment configuration.
Accessibility	Limited to a single platform-specific device	Allows remote access from any device with a browser and internet connection.

the Python standard library and other modules. It is responsible for loading Python code and executing it stepping through each line.

An interpreter operates like a command line shell reading individual typed (or pasted) lines and executing them. It is in interactive mode when it prompts for the next command, usually with three greater-than signs >>>. An example online interpreter available at https://pyodide.org/en/stable/console.html is shown in Figure 0.1. The Pyodide console requires no account and can be the most rapid method to access Python. It is also open source and customizable (the author has restyled it, added hotkey operations, and posted it on web servers).

The interpreter session in Figure 0.1 demonstrates five statements starting with the print command on the first line followed by its output, the setting of variables, a variable equation, and an evaluation of the variable that prints the value.

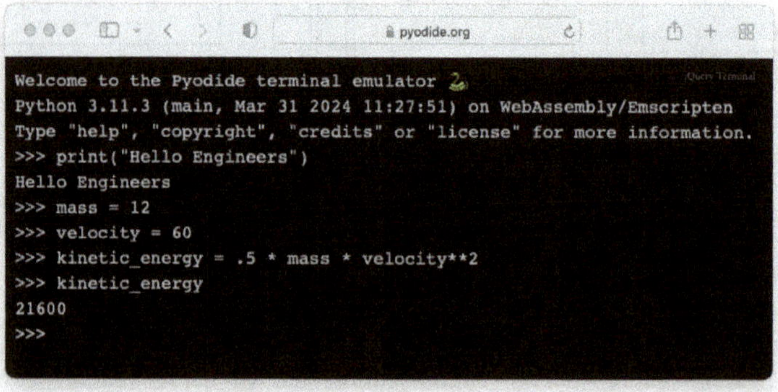

Figure 0.1   Online Interpreter with Pyodide

The interpreter command line mode is also called a Read-Evaluate-Print-Loop (REPL) environment. Each line is read, evaluated, and the results optionally printed (all expressions are printed), and the interpreter loops again for the next input command. Print statements are not necessary in the REPL to display individual expression values, because it is built in that all expression evaluations are automatically printed.

An interpreter is handy for one-off calculations, rapid prototyping, testing code snippets before putting them into larger programs, getting help, or exploring language features. The statements are not saved to a file, so an editor is necessary for that.

0.2.2   IPYTHON

IPython is an advanced Python shell for interactive computing. It enhances the standard Python interpreter with additional features such as syntax highlighting, code completion, easy access to shell commands, and an extensive history mechanism.

IPython is designed to optimize the user experience with a more informative and responsive command-line interface (see further details and examples in Chapter 5).

IPython is the foundation for sophisticated IDEs described later. These IDEs extend IPython by integrating it with a web-based interface for combining code execution, text, and multimedia content in a single, interactive document. It is included in the desktop Anaconda distribution IDEs Jupyter Notebook, JupyterLab, and Spyder for enhanced interactive computing. Visual Studio Code does not come with IPython, but it supports Jupyter Notebooks, and IPython can be installed separately. Cloud-based IDEs based on IPython include Google Colab, Anaconda Cloud, Microsoft Azure Notebooks, Kaggle Kernels, Binder, and others.

0.2.3 JUPYTER NOTEBOOKS

A Jupyter Notebook contains integrated code and text cells functioning as an interactive engineering lab notebook. It combines code, formatted text, graphics, tables, and other visualizations in a single document which can be exported to different formats. They are useful for prototyping, code development, data exploration, analysis, and visualization, report generation, and training. Cloud-based IDEs enable notebook collaboration with multiple people.

A Jupyter notebook code cell may contain a full program or a single command and work like a traditional interpreter. The output fields behave like traditional console output where user input takes place. An example notebook is shown in Figure 0.2 in the JupyterLab IDE. Jupyter Notebook is both a separate tool and the name of the open source format for Jupyter notebooks supported by other IDEs (files with an .ipynb extension).

Figure 0.2 shows after execution that the code cells are numbered (by most recent execution order), and when executed the printed output goes below. Each cell may contain full scripts or perform as an interpreter for single lines. New code cells are generated automatically at the end.

0.2.4 LOCAL DESKTOP ENVIRONMENTS

A Python interpreter, which will be necessary, may already be installed. This can be determined by opening Powershell on Windows, or the Terminal application on MacOS or Linux, and typing "python" at the command prompt. If Python exists, it will display its version and go into an interpreter mode per the top of Figure 0.3. If the version shown is 2.x, then try "python3" instead. If it is not installed, choose one of the tool options described next.

Local interpreters are invoked in command shells for PowerShell and a Terminal window in Figure 0.3 with the `python` command. After typing expressions or print statements, they respond and add another command prompt waiting for the next Python statement. Local interpreters work identically with MacOS Unix, Windows Powershell, or a Linux terminal.

A local interpreter can also read and execute a script from a file when called with a file name or with a file as standard input. The latter happens in a Python

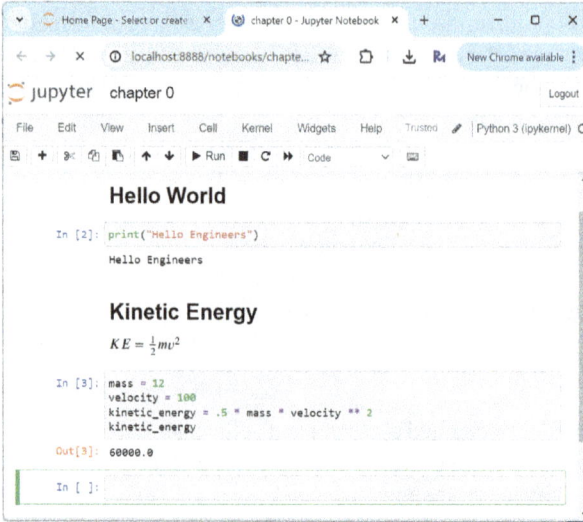

Figure 0.2   Jupyter Notebook with Text and Code Cells

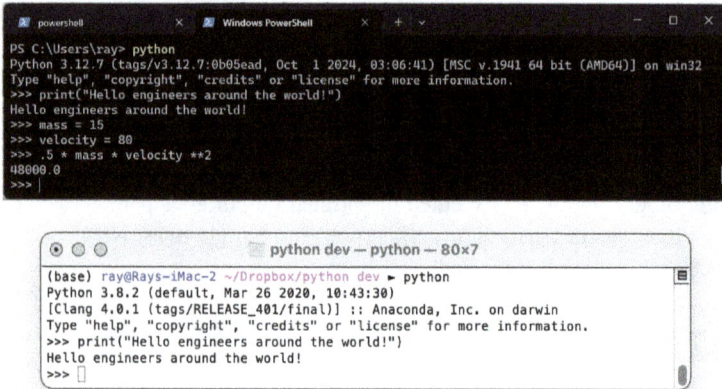

Figure 0.3   Interpreters in PowerShell and Terminal

IDE when given a run command for the program currently in the editor. Figure 0.4 from a terminal window shows how a Python script with a .py extension can be sent directly to the interpreter via the command line to be executed. The referenced program *hello_engineers.py* consists of the print statement.

Beyond the Python interpreter itself, one will need a text editor at minimum. It is best to use a more comprehensive tool environment like Visual Studio, Anaconda

```
(base) C:\Users\ray>python hello_engineers.py
Hello Engineers Around the World!

(base) C:\Users\ray>_
```

Figure 0.4   Program Execution in Terminal via Python Command

Spyder, JupyterLabs, Jupyter Notebook, or other IDE. Desktop IDE's can be highly customizable. They allow installation of any custom packages, IDE configuration, and setting up of alternate environments.

A desktop Python deployment is diagrammed in Table 0.1 requiring a Python interpreter running under the local operating system to read and execute local files. Everything runs on local hardware without the need for an internet connection. The files may be individual python files (*.py) or Jupyter Notebook files (*.ipynb). Files are managed using the operating system's file system, and performance is dependent on local hardware resources (CPU, RAM, etc.).

For engineering applications on the desktop, it is recommended to use the Anaconda Python distribution that comes with standard scientific libraries and development tools including the the Scientific Python Development Environment (Spyder), Jupyter Notebook, Jupyter Labs with additional utilities beyond Jupyter Notebook, and other programs. It will also install Python.

The Spyder editor comes with an IDE specifically designed for scientific computing, as shown in Figure 0.5. It contains sophisticated features such as code completion, a variable explorer, integrated documentation, and debugging capabilities. Figure 0.5 shows a program script in the editor on the left, and execution on the right side interpreter console with its output. User inputs are also provided in the console, and it can be used independently as an interactive shell. The interpreter console shows the default IPython numbered prompts **In [#]:** and **Out [#]:**. Not shown are several other window views available in Spyder.

The Visual Studio IDE is shown in Figure 0.6. The script on the top pane is executed in the interpreter window below it. Multiple interpreter tabs are shown open in this example, each of which can be executed separately at the command line.

Though a comprehensive IDE is recommended, a more basic Python installation can be downloaded from https://www.python.org/ [4]. It also includes the IDLE editor, which is more primitive compared to other tools. It will be necessary to manually install additional commonly used libraries, e.g., NumPy and Matplotlib. One can be more productive and capable with a modern, sophisticated Python tool environment like the recommended Anaconda distribution for local operations or a cloud-based IDE.

0.2.5   CLOUD-BASED ENVIRONMENTS

A cloud-based development environment can be the easiest to start with since no local installation is required. However, an account is necessary to save files. These IDEs may be suitable if there are installation constraints, or necessary when away

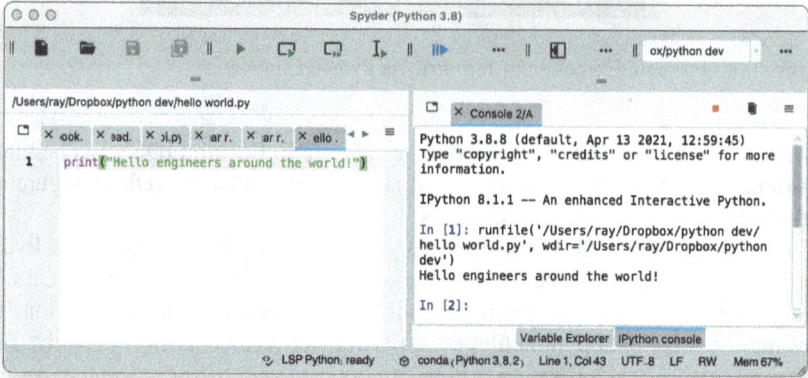

Figure 0.5   Spyder IDE Editor and Interpreter Console

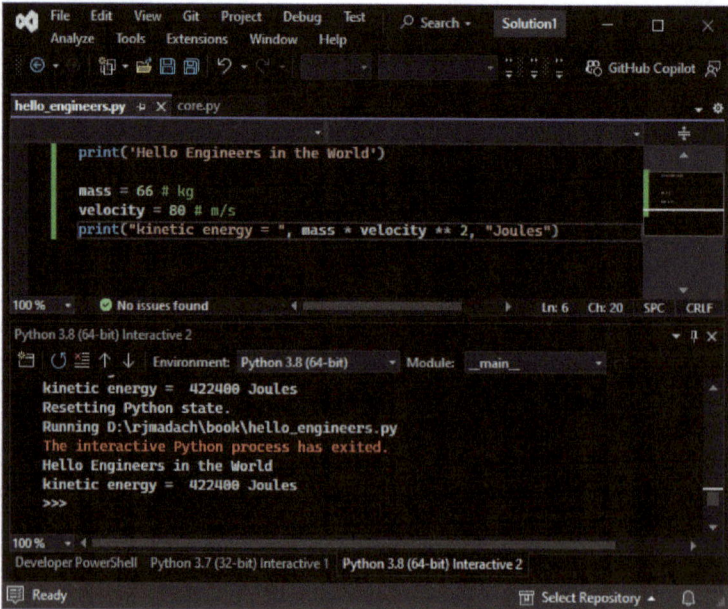

Figure 0.6   Visual Studio IDE

from one's computer(s). These environments always depend on an Internet connection, and may have processing limitations.

The deployment diagram in Table 0.1 shows the topology of a cloud-based platform. A user interacts through a local browser connected via the Internet (using the HTTP protocol) to a web server running the Python interpreter. Files are typically located in a corresponding cloud account (or uploaded for a session) to be executed. Cloud-based environments include Anaconda Cloud, Microsoft Visual Studio, Google Colaboratory (Colab), and others. Most of these support Jupyter Notebooks.

Python and major libraries are pre-installed on the cloud servers. Files being developed (e.g., *.py or *.ipynb) are saved in the cloud to connected storage. Uploaded or generated files are temporary during a virtual session and must be explicitly saved. Datasets, outputs, or logs must be manually saved to avoid loss after a session.

These environments are ideal for remote access and working on multiple devices without needing to install anything. Another advantage of cloud platforms is real-time collaboration, allowing multiple users to work on the same files simultaneously.

However, cloud-based platforms have limited customization. Package installation and environment configuration may be restricted by the cloud service, and custom libraries may need to be installed in each session. Performance depends on the cloud service's computing resources.

Colab is a free cloud-based platform built on open-source IPython, integrated with Google Drive. An example Colab notebook with code and text is shown in Figure 0.7. It has an additional table of contents and other features enhancing the standard Jupyter Notebook format.

Another example IDE is at `https://replit.com`, where a free account is required to save and share work. The Replit cloud-based editor in Figure 0.8 shows the program in the left pane and its execution in the right console/interpreter (which can also be used independently to enter and run statements). Currently, it does not support Jupyter Notebooks.

One can also embed REPLs on local or internet web pages without requiring an account. The open-source tool *PyScript* provides a browser-based interpreter in HTML files. An open source JavaScript implementation of a Python interpreter using Pyodide is instantiated on the web page. This method provides code syntax coloring but currently doesn't have smart completion features within the editor like other IDEs. See Section 4.2.4 for more detail.

0.3 INTRODUCTORY PYTHON SYNTAX AND FEATURES

Certain characters in Python syntax have special interpretations by the interpreter during execution. One example was the use of double quotes in the *hello world* script to delineate the character string for printing. An overview of introductory syntax covered in this chapter is in Table 0.2 for assignment statements, strings, lists, and control statements.

In Python, *keywords* are reserved words used by the interpreter to parse a program. This chapter introduces the `for` and `in` keywords. Python *built-in functions* are pre-defined and available for use as part of the Python standard library, such as

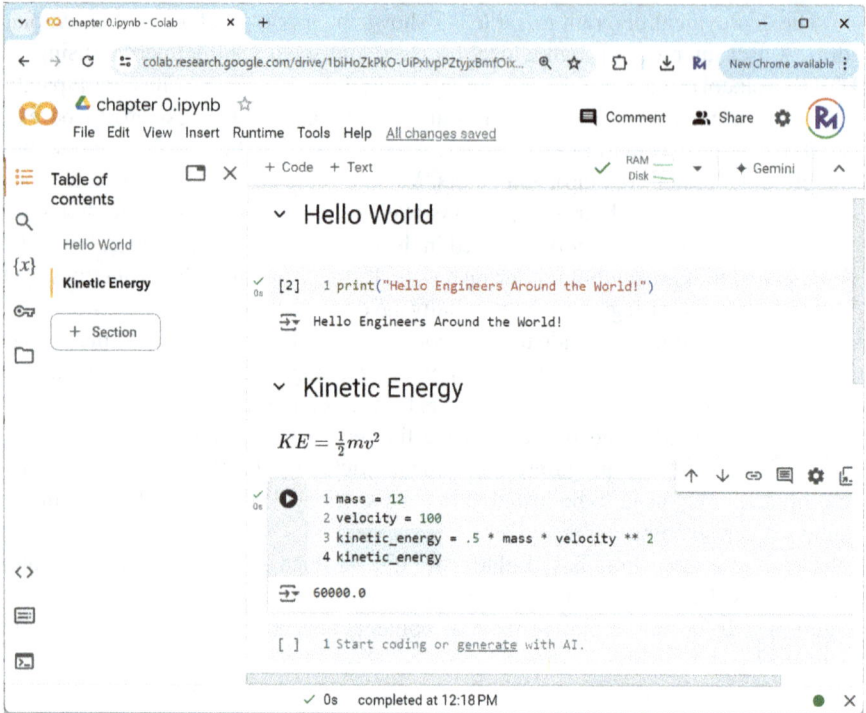

Figure 0.7   Google Colab Notebook

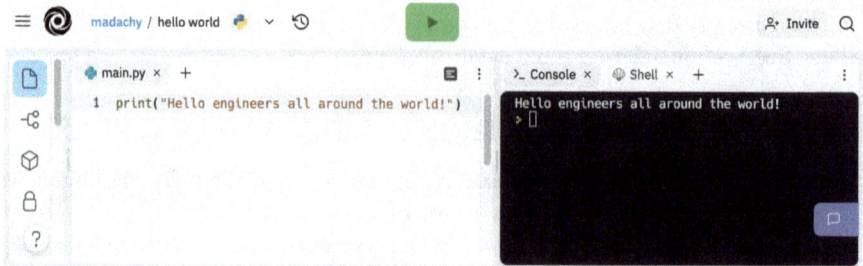

Figure 0.8   Cloud-Based Editor at https://replit.com

Table 0.2

Introductory Python Syntax

Characters	Description and Examples

=	Assignment operator used to assign a value to a variable.
	
	 `reliability = .98`
	`message = "Hello engineers."`
	`measurements = [67.8, 73.1, 79.3, 81.4, 69.9]`

' '   " "
 ''' '''
 """ """
	Character strings are surrounded by single, double, or triple quotes such as `"Hello engineers."` or `'battery'`. A string containing a quote character must be delimited by another quote type such as `"Ohm's Law"`.
	 Triple quotes are sequences of three single quotes or three double quotes, and can be used for multiline strings with line endings.
	
	 `pyscript_code = """`
	`<py-repl auto-generate="true">`
	`print("Hello engineers around the world!")`
	`</py-repl>`
	`"""`

[]	Lists are sequences of items separated by commas surrounded by square brackets. Lists may contain numbers, strings, mixed data types, other lists, and other entire data structures. `['transmitter', 'battery', 'antennae'] [2, 4, 5] [[62, 6↙ 4, 61], [60, 61, 59], [61, 60, 64]]`

:	A control statement has a colon at the end which denotes that a nested block of statements follows next. The block may be iterated over with `for` or `while` control statements as below, or conditionally executed with `if`, `elif`, and `else` statements per Chapter 1.
	
	 `for velocity in [25, 50, 75, 100]:`
	`    height = (velocity**2)/g/2.`

---

the `print` and `range` functions demonstrated. Appendix A Tables A.1 and A.2 have complete lists of Python keywords and built-in functions.

The syntax characters are invariant across IDEs, but styles for coloring and fonts can vary across environments, and normally several styles are available to choose from. The coloring scheme in this book matches the Spyder IDE light color scheme default. This scheme shows keywords in blue (`for`), built-in functions in magenta (`range`), character strings in green (`'weight'`) and numbers in dark red (`99`). All source code examples and program output are displayed in courier text which is conventional for code and the default Python interpreter font.

Other basic Python constructs are introduced next before a fuller treatment in Chapter 1:

- Variables store data values, and they are assigned with the = operator.

- A *list* data structure is an iterable sequence of values contained in square brackets separated by commas such as `[1, 2, 4, 8, 16]`.

- Data structure indices start at zero (0) as exemplified in this Chapter number.

- Computation loops can be controlled with a `for` statement that iterates across a sequence.

- Comprehensions are powerful single line statements of `for` loops demonstrating the Pythonic way, e.g. `squares = [num * num for num in range(100)]`.

0.3.1 VARIABLES AND ASSIGNMENT STATEMENTS

Variables are used to store data values. A variable can be assigned any valid Python data type, such as numbers, strings, and more complex data structures like lists. Assignment is done using the = operator, where the variable on the left-hand side is assigned the value on the right-hand side. The examples below assign values for a floating point number, integer, and string respectively. These variables are assigned *literal* values because they take on the data type as explicitly written.

```python
safety_factor = 3.5
number_tanks = 12
material = "tungsten"
```

0.3.2 LISTS

Lists are convenient general-purpose data objects containing sequences of data values in brackets separated by commas. The assignment statement below assigns a list of numbers to the variable `angles`. The list structure is an *iterable* that can be looped through to perform calculations for each value in it.

```python
angles = [0, 30, 60, 90]
```

List elements may be numbers, strings, mixed data types, other lists, or other entire data structures. Lists can be populated programmatically in various ways, including through list comprehensions as shown later.

0.3.3 ZERO-BASED INDEXING

Python uses zero-based indexing of data sequences, like many other languages including Java and C. Starting with an index of zero instead of one may initially seem counter-intuitive, so this Chapter 0 will serve as a reminder. Given the following list of this book's chapters, the initial value is specified with an index of 0:

```
chapters = ["Introduction", "Language Overview", "General Purpose ↙
    Scientific and Utility Libraries", "Engineering Analysis ↙
    Examples", "Python Applications", "Processes and Tools"]
print("We are in Chapter 0", chapters[0])
```
```
We are in Chapter 0 Introduction
```

Lists and other data structures are also indexed starting at zero. Functions like range that provide counting sequences start at zero. These constructs are shown next for controlling loops.

0.3.4 FOR LOOP

A for statement is a control statement used to iterate over the elements of a sequence (e.g., a list or other iterable object) per the loop syntax below. The nested statements will be repeated as a block for each item in the sequence, where the variable name will take on the value of each item. In the statement both for and in are highlighted as keywords shown in blue, as is the case in most environments.

For Loop Syntax

```
for variable in sequence:
    # block of code to be executed for each element in sequence
```

One way to control the loop is with a list:

```
for part in ["controller", "frame", "battery"]:
    print(part)
```
```
controller
frame
battery
```

The range() function is often used in a for statement to control the looping. It returns a sequence of numbers starting from 0, incrementing by 1, up to but excluding the stop number. The minimum input it takes is a single argument for the stop number as below.

```
for car in range(3):
    print("Car number", car)
Car number 0
Car number 1
Car number 2
```

0.3.5 COMPREHENSIONS

Lists and other iterable sequences can be computed via a set of looping and filtering instructions called a *comprehension*. A comprehension is a handy shortcut method for creating and populating sequences. It consists of a single expression followed by at least one `for` statement clause and and optionally includes one or more `if` clauses (conditional statements introduced in Chapter 1). The basic form of a list comprehension is shown below. The square bracket syntax indicates a list is to be created. Comprehensions for other iterable data structures are similar except for having different outer brackets per Chapter 1.

List Comprehension Syntax

```
[expression for variable in sequence]
```

Next, a larger list of angles compared to Section 0.3.2 is generated with an automated list comprehension. Using the `range` function the list will start at zero degrees, increment by 10 up through 90. Incrementing up to and not including 91 produces equally spaced intervals when considering the interval boundaries (another advantage of zero-based indexing).

```
angles =[angle for angle in range(0, 91, 10)]
print(angles)
[0, 10, 20, 30, 40, 50, 60, 70, 80, 90]
```

The above comprehension applies no operations to the `range` output, but the expression can be any degree of complexity. For example, radians are often used in geometry computations and the required unit for scientific libraries, so the angle expression will be modified for the conversion *radians = degrees* $* \pi/180$. If π is previously defined in a library or a code statement like `pi = 3.14159`, the following will generate radian values. Alternatively its numerical value could be used instead of the `pi` variable name.

```
radian_angles = [degree * pi/180 for degree in range(0, 91, 10)]
print(radian_angles)
[0.0, 0.1745327777777778, 0.3490655555555556, 0.5235983333333333,
0.6981311111111111, 0.8726638888888889, 1.0471966666666666,
1.2217294444444444, 1.3962622222222223, 1.570795]
```

The resulting values can now be input directly to a math function for computation with radians. As a best practice, the angle variable names were changed to be more explicit and to avoid any unit misinterpretations.

More complete details and examples of these introductory features are in Chapter 1 and many applications in subsequent chapters. The next example collectively demonstrates these introductory features as applied to a simple engineering problem. It will be further elaborated to take advantage of additional language features and Python scientific libraries.

0.4   INITIAL PROJECTILE MOTION EXAMPLE

The analysis of projectile motion is a recurring example throughout the book. These analyses are applied for the fictitious company ProLaunch that designs projectile launchers for scientific, educational, and humanitarian aid purposes. Initially it is desired to calculate the maximum projectile height as a function of the initial velocity when launched perpendicular to the ground. Figure 0.9 illustrates the geometry.

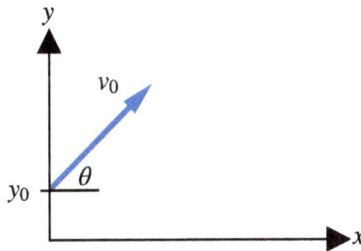

Figure 0.9   Launcher Geometry

From the equations of motion, the maximum height H of a projectile with an initial velocity v_0 having v_x and v_y components when $y_0 = 0$ is:

$$H = \frac{v_y^2}{2g}$$

where g is the gravity constant.

It is desired to evaluate a launcher capable of up to 100 m/hour velocity in increments of 25 m/hour. Accordingly the height calculation will be repeated for each desired velocity. This is the essential logic of a for loop. The program below computes and prints the height for four values of velocity contained in a list. The first line is an *assignment statement* to initialize the gravity constant. The next line is a *for control flow statement* to control the underlying computation loop which is repeated for each value of velocity.

```
g = 9.8# meters per second
print('velocity height')
for velocity in [25, 50, 75, 100]:
    height = (velocity**2)/g/2.
    print(velocity, height)
```

```
velocity height
25 31.887755102040813
50 127.55102040816325
75 286.98979591836735
100 510.204081632653
```

The output of velocity and height pairs could be used to populate a more complete table or graph for summarizing the analysis. The above `for` loop for height computation can be collapsed into a single line *list comprehension* to create a list of lists. The next example uses finer gradations of velocity and an automated way to specify the velocities in a sequence without writing each one manually.

The outer list `velocity_heights` is created below with an element for every 10 m/s division of velocity. Each element is a list containing two values for velocity and height. Since list indices start at zero, the parameters can be identified with the values 0 for velocity and 1 for height in the print statement.

```
velocity_heights = [[velocity, velocity**2/g/2] for velocity in ↵
    range(0, 101, 25)]
print('velocity height')
for data_point in velocity_heights:
    print(data_point[0], data_point[1])

velocity height
0 0.0
25 31.887755102040813
50 127.55102040816325
75 286.98979591836735
100 510.204081632653
```

In the comprehension square bracket syntax was used to define the main outer list and within that to define the contained lists of two values for each outer element.

In subsequent chapters, this example will be progressively augmented for more complex geometry, launch conditions including air resistance, and fine tuning of the formatted output. The parameter set of the problem will be increased, and accordingly we'll learn alternative ways to specify them in code with more convenient methods. *Dictionaries* are alternative data structures to lists that use keywords for this purpose as described in Chapter 1.

0.5 PYTHON BACKGROUND

Python is shown in the historical timeline of major languages in Figure 0.10 being introduced in 1991 (this timeline was generated with Python). It has borrowed from predecessor languages, improved upon them, and also contributed to developments in other languages and computing platforms. Engineering and science applications have been major drivers of computing technology advances, starting with the first programming language FORTRAN which was developed for engineers. Python is a bridge between older, lower-level languages like Fortran, C, and C++ and modern languages designed for speed and efficiency, such as Rust and Go.

Python's emphasis on readability, modularity, and ease of use was a direct response to the complexity of earlier languages like C. Although implemented in C, Python abstracts away much of its complexity through high-level constructs. It was also part of a broader movement away from procedural languages toward object-oriented languages designed for solving complex engineering problems with reusable, modular code.

Guido van Rossum created and first released Python in 1991, focusing on code readability and simplicity. It has evolved with critical improvements in functionality, performance, and maintainability. Its importance in scientific computing was highlighted by Oliphant in a seminal article [15] about the language's potential for scientific applications. It has since become a standard tool in engineering, driven by its extensive ecosystem of libraries and tools, ease of use, and strong community support.

In parallel, IPython was created by Fernando Pérez in 2001 as an enhanced interactive shell for Python to improve its usability for scientific computing. It introduced features including advanced auto-completion, better history tracking, and inline visualization. IPython expanded to support Jupyter Notebooks in 2011. Its open-source nature has further contributed to its evolution into the broader Jupyter ecosystem, which now supports many other languages alongside Python. R is another notable example for scientific computing commonly used with the Jupyter ecosystem.

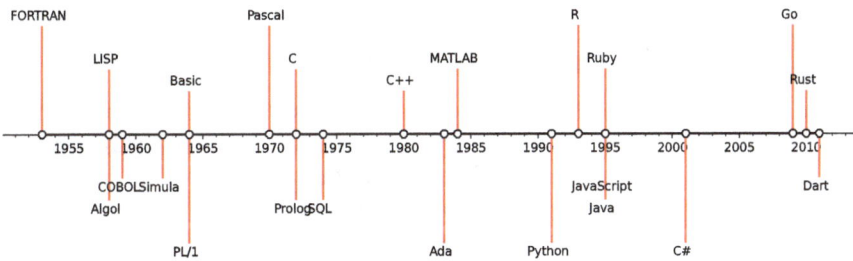

Figure 0.10   Major Programming Languages Timeline

The Figure 0.10 timeline also shows newer languages, many of which are frequently integrated with Python. Multiple languages may be used for a particular engineering or science application. JavaScript is a good example because it is used in conjunction with Python to support responsive web pages in a browser (see Chapter 4).

Python has been ranked the #1 language for engineers the last 6 years running per the Institute of Electrical and Electronics Engineers (IEEE). IEEE is the world's largest technical professional society and annually publishes the top programming languages for engineers. Recent rankings in Figure 0.11 [2] are based on multiple weighted criteria. The figure shows the top 15 by different rankings. The *IEEE Spectrum* ranking is heavily weighted toward the interests of IEEE members, while *Trending* puts more weight on forums and social-media metrics. Per IEEE Spectrum's more technical criteria, Python is followed by C variants and Java which are far above all the others.

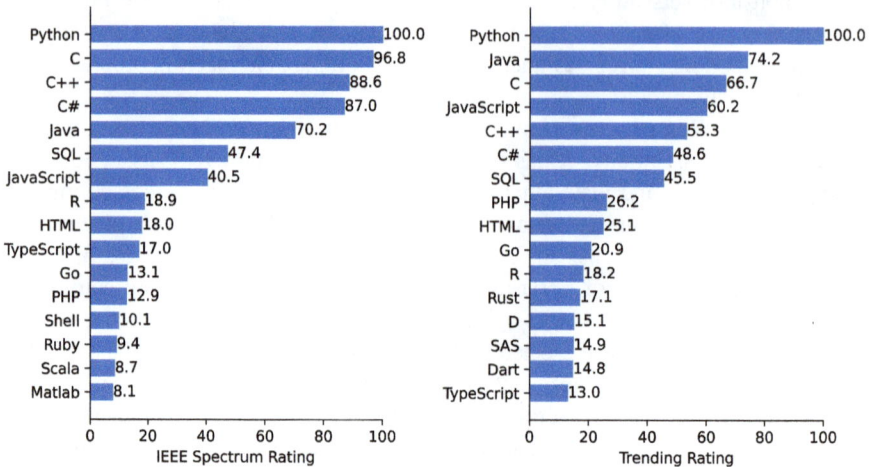

Figure 0.11   Top Programming Languages 2022 – From *IEEE Spectrum*

0.5.1   SUPPLEMENTARY REFERENCES

There are additional valuable references for engineers, ranging from starting problems to extensive engineering analysis applications. A more thorough introduction to basic programming concepts is in [3]. It provides guidance on constructing small to medium size programs for physics and engineering with Python (though the demonstrated tools are now outdated). For a data-centric approach to scientific computing, [5] focuses on data science applications, but it lacks coverage of physical engineering problems.

Many references concentrate on advanced applications for numerical analysis. For example, [7] presents comprehensive examples suitable for upper-division or graduate courses, focusing on numerical methods in engineering. Another resource for scientific programming with a wide spectrum of applications is [9]. It has a significant emphasis on algorithms.

The book [8] provides in-depth language definition details, with examples of generic mathematics algorithms. Its approach involves writing algorithms for a narrow class of problems from scratch without using open source. For applied mathematics, [6] offers detailed explorations of numerical computing and statistical topics, such as solving sparse matrices. However, the examples are generic and lack context for specific engineering applications.

The popular book [13] provides a thorough coverage of using open-source libraries for data analysis to supplement Chapter 1. Using Python to support workflows, as discussed in Chapter 5 is well covered in [21]. It emphasizes Python's role in automating repetitive engineering tasks such as data processing, file handling, and testing scripts.

0.6 SUMMARY

A Python program consists of a series of statements that are executed sequentially by an interpreter. Examples include print, assignment, and control statements. Assignment statements define variable values, while control statements are used for conditional execution. Python programs can be run in various environments, including command-line interfaces and IDEs.

The Python interpreter provides an interactive environment for executing code one line at a time, making it useful for rapid prototyping, testing, and debugging. It operates as a Read-Evaluate-Print Loop (REPL), automatically printing the results of expressions typed into it.

IDEs are integrated platforms for writing, editing, and running Python scripts. These environments include features for code completion, variable exploration, integrated documentation, and debugging capabilities. IDEs may be desktop-based, where all files are managed locally, or cloud-based, offering remote access and collaboration features.

IPython is an open-source foundation for sophisticated IDEs. Jupyter Notebook, a tool and file format based on IPython, allows users to combine code, text, tables, graphics, and other visualizations in a single document. These are especially useful for data analysis, scientific computing, and real-time collaboration.

Local desktop IDEs include Visual Studio, Anaconda, Spyder, Jupyter Notebook, and JupyterLab. These IDEs offer flexibility, allowing installation of custom libraries and packages, without the need for an internet connection. Cloud-based IDEs like Google Colaboratory, Anaconda Cloud, and Replit require no local installation and support online collaboration.

Key constructs in Python syntax include variables, lists, and control statements. Variables store values and are assigned using the = operator, while lists hold sequences of data. Control flow is managed using conditional statements and loops. Lists can be generated concisely with single-line comprehensions or processed in loops governed by control statements.

Historically, Python was created to improve readability, modularity, and ease of use. It has borrowed features from predecessor languages, improved upon them, and contributed to advances in other languages and computing platforms. Python is now the dominant language in scientific and engineering fields due to its extensive ecosystem of libraries, ease of use, and strong community support.

LANGUAGE COMPARISONS

Python is executed directly in an interpreter, as contrasted with languages like Java and C which require extra compilation steps to transform the source code into an executable form. The translation is thus simpler but most gains using Python come from its superior readability. Code examples from Java and C will help illustrate structural and readability advantages of Python.

Python code is made sparse by eliminating superfluous and distracting characters for logical nesting and statement ends. In Java and C (as well as their derivatives like C++ and C#), additional syntactic elements are used to structure code. These include the use of braces to define blocks of code and semicolons ; to end statements. The blocks can be aligned anywhere without any visual nesting. These add to the complexity of reading the code and potentially lead to errors. In contrast, Python utilizes indentation to structure code blocks, making it inherently more readable.

Furthermore, Python's syntax is less verbose. For instance, to print a simple message like "Hello," Java requires the declaration of a class with a main method, including various keywords and punctuations, as shown below:

```
class HelloWorld {
    public static void main(String[] args) {
        System.out.println("Hello engineers around the world!");
    }
}
```

C also requires extra overhead. A named main program must be created; blank parentheses to denote no inputs, brackets for the block structure, and a semicolon line ending are always required. Here the "stdio.h" is an external library file included for standard input and output.

```
#include <stdio.h>
int main()
{
    printf("Hello engineers around the world!");
}
```

The equivalent Python code is just a single line, without the need for additional structural syntax.

The other languages will not be going away, and Python must co-exist with them. It behooves the engineer to understand how they work in general, and in order to compare language alternatives. For example, if execution speed is the primary concern, then a compiled language may be faster than Python due to compiler optimizations.

0.7   GLOSSARY

assignment:: A statement that assigns a value to a variable.

argument: A value passed to a function (or method) when calling it.

comprehension: A concise way to create lists or other sequences by defining an expression followed by for and if clauses.

expression: A combination of variables, operators, and values that produce a single result.

iterable: An object capable of returning its members one at a time. Iterables include all data types containing sequences (including lists) and other non-sequence data types. Iterables can be used in a `for` loop and other places where a sequence is needed.

keyword: A reserved word that is used by the interpreter to parse a program. Keywords are in Appendix Table A.1.

list: An ordered collection of items. Lists are defined by square brackets [] and can store elements of different data types.

statement: An instruction that executes a command or action including assignment statements and control statements.

1 Language Overview

1.1 STYLE AND CONVENTIONS

A driving goal for Python is to enable concise and highly readable code. It achieves this through visual code structure and its language features. The code layout rules enforce legible, readable code through indentation to delineate logical blocks of code with no extraneous characters. Statements simply end with a return key [1].

Highly readable code that is self evident and well documented makes for easier, quicker, and more reliable changes when inevitably modifying the code later. The term *Pythonic* refers to a program generally as short as possible, but not at the expense of legibility.

1.1.1 LAYOUT, INDENTATION, AND WHITE SPACE

Python relies on indentation, whitespace at the beginning of a line, to define scope in the code and make the logic nesting evident. This is enforced syntax for programs to run. Below are two representative examples of how indentation marks logical structure. Upon inspection, the indentation makes clear to the reader what the program logic is doing before knowing the language details. In these examples, each logical block is prefaced above with a statement ending with a colon (:). Smart Python editors will automatically indent the blocks when typing or pasting statement lines.

```python
# indentation of logical condition blocks
if weight > weight_max:
    print("Weight exceeds the maximum.")
else:
    print("Weight is within limits.")


# indentation to define a function
def kinetic_energy(mass, velocity):
    kinetic_energy = .5 * mass * velocity * velocity
    return kinetic_energy

print(kinetic_energy(100, 10))
```

The first example shows indented conditional blocks of code using the if statement described later in Section 1.5.1. There are two logical paths based on the value of weight to print the result. It is clear in the second example that the function computes kinetic energy as $\frac{1}{2}mv^2$ in the indented lines.

The function definition for kinetic_energy takes inputs for mass and velocity and returns the kinetic energy. It is prefaced with the def keyword and ends with a

---

[1] Optionally one can add a semi-colon ; to designate end of statement in order to put multiple statements on one line. This is not recommended except for special cases to conserve lines.

colon (:), indicating a following block of code. The indented code in the function is run each time it is called (see Section 1.2 on functions).

Some example language features described later that make for powerful and elegant code include: iterable data structures where looping and global operations are automated (e.g., sequences including lists and dictionaries), single line comprehensions to create data structures using `for` loops and logical `if` conditions, nesting of expressions, passing of functions to other functions, handy functions to operate on data structures, and more.

1.1.2 VARIABLE NAMES, ASSIGNMENTS, AND DATA TYPES

Variables are named containers for storing data values. The names of variables, functions, and classes are all case-sensitive. Variable names are conventionally written in *snake case* using lower case letters with spaces between words replaced with underscore (_) characters, e.g. `kinetic_energy`. The name visually invokes a snake between words. *Camel case* has new words starting with uppercase letters, e.g, `electricCar`, resembling the hump of a camel and is sometimes used for naming Python classes.

Uppercase characters are frequently used for variable names of fixed constants that won't vary during execution, such as for physical parameters. This practice isn't universal as is demonstrated in examples throughout this book.

An assignment statement uses the equals sign = to assign values to variables, as well as data structures containing collections of values (e.g. lists).

```
number_robots = 14
material_type = "tungsten"
GRAVITY = 9.8
potential_energy = mass * GRAVITY * height
pressure_settings = [220, 240, 260, 280, 300]
```

These statements define the variables `number_robots`, `material_type`, and `GRAVITY` to contain integer, string, and floating point values respectfully as described next. Additional shortcut assignment statements are detailed in Section 1.4.4.

Different types of variables include the following:

- **Integers** are whole numbers like 714

- **Floating-point numbers** contain a decimal point like 3.14

- **Strings** are text as sequences of characters enclosed in single, double, or triple quotes like `"Hello Engineers"`

- **Booleans** are one of the two logical values `True` or `False`

- **Literals** are explicit values written out by themselves to represent any of the above types. The types are determined by the values provided, hence "literal". E.g., the previous variables `number_robots`, `material_type`, and `GRAVITY` were defined respectively as integer, string, and floating point as explicitly typed.

1.1.3   STRINGS

A string is a sequence of characters enclosed in single quotes: '...', double quotes: "...", or triple quotes (of a single or double quote symbol). A string enclosed in single quotes may contain double quotes, and vice versa.

Triple quotes can be used to write strings that extend over multiple lines when each line is significant. Situations for multiple lines include comments, data streams, or program statements of another language to interpret. The triple quotes method can also be used for multi-line comments and code *docstrings* per Section 1.14.3.

An example of a string on multiple lines is below. This example with """ is equivalent to using triple quotes with ''':

```
test_data = """
60, 63, 55, 67
46, 56, 35, 74
16, 83, 35, 27
"""
```

In the above each line may represent a data point of multiple values. Another use case for triple quote strings is writing multi-line files without resorting to specially escaped line ending characters (see below).

Some special characters cannot easily be entered directly into strings and must be "escaped" using a backslash character (\) per:

\n represents a newline character

\t represents a tab character

\' represents a single quote (inside a singly-quoted string)

\" represents a double quote (inside a doubly-quoted string)

Strings can be concatenated, or joined together, with the + operator.

```
print("Hello engineers " + "around the world.")
Hello engineers around the world.
```

1.1.4   NUMBERS

There are numeric types for integers, floating point numbers, and complex numbers. Numbers are created by numeric literals or as the result of built-in functions and operators. The functions int(), float(), and complex() can be used to produce numbers of a specific type (also called *casting*). Booleans are considered a subtype of integers. Complex numbers have a real and imaginary part, both of which are floating point numbers.

1.1.5   LITERALS

A literal is an explicit value written out by itself as shown in Section 1.1.2 to represent strings, numbers, Booleans, larger data structures (e.g. lists and dictionaries), or special literals. A string literal is surrounded by single, double, or triple quotes. Numeric literals are integers written without a decimal point, a floating point number with a decimal point, or other numeric system (binary, octal, hex, etc.). The values and types of the literal variables can be changed in later statements.

1.1.6   VARIABLE CASTING

Variables do not need to be declared as a particular type as in other languages, but can be done with casting. This would be necessary when converting from a string to a floating point number using `float`, such as when receiving user input as a string to convert for calculations as in Section 1.11). Sometimes discretization is necessary to convert floating points to integer values with `int`.

```
>>> str(3.14159)
'3.14159'
>>> int(3.14159)
3
>>> float('3.14159')
3.14159
```

1.2   FUNCTIONS

A Python *function* is a reusable named piece of code that can return value(s) or perform other operations with given input. Functions are fundamental building blocks in programming and are a best practice to encapsulate frequently used code in a single place, making the code more modular and manageable. When an algorithm, or a set of statements, needs to be executed for multiple inputs or in various sections of a program, it is inefficient and error-prone to repeat the code each time. This not only makes it difficult and time-consuming to read but also complicates maintenance, as any updates or bug fixes would need to be applied in multiple locations manually.

A function, also called a subroutine, can be written once and called many times with different parameterized inputs. This modular approach saves program space, improving readability, and also allows one to better understand the overall structure by breaking it down into logical components. Functions make testing and debugging easier, as errors can be isolated to specific code blocks rather than scattered across the program. These practices collectively reduce errors and enhance code quality, which is especially important in larger programs.

The syntax for a function is shown next. The keyword `def` introduces a function definition. It is followed by the function name and a parenthesized list of comma-separated input parameters. The colon at the end indicates that the following indented lines form the function code block. The indentation helps maintain a clear structure.

Function Syntax

```
def function_name(parameters separated by commas):
    """Optional docstring describing the function"""
    # Set of statements including optional return() statement to ↙
        send back value(s)
```

A function can be called for execution by giving its name and any arguments it requires in parentheses. The arguments are matched up to expected input parameters. There must be the same number of calling arguments as defined parameters, and they are matched by position (except when using keywords per Section 1.2.2). A function may or may not return value(s). If a function only prints output for example, there is no explicit returned value. Multiple values can be returned and can be of different data types.

For a very simple function requiring no parameters (e.g., print time of day) the parentheses are left blank. A function that returns value(s) will have a `return()` statement with returned value(s) contained in its parentheses. The optional *docstring* at the beginning in triple quotes describes the function and how to use it. The syntax supports automated help and usage documentation, so it's recommended to include docstrings (see later demonstrations).

An example function for calculating the volume of a sphere given the radius can be defined as:

```
def sphere_volume(radius):
    """ Calculates the volume of a sphere given its radius. """
    PI = 3.14159
    volume = 4 / 3 * PI * radius ** 3
    return(volume)
```

Calling the function can be done with the code `volume(radius)` and the calculated volume will be returned. Below the call is embedded within a print statement, which is covered in Section 1.3. It demonstrates printing an *f-string*, as prefixed with the `f`, which is covered in Section 1.9.1.

```
radius = 5
volume = sphere_volume(radius)
print(f'Volume of a sphere with radius {radius} is {volume:.1f}')

Volume of a sphere with radius 5 is 523.6
```

The projectile motion example will be updated as a function below. Given the initial velocity and launch angle, the function will return the flight time, maximum height, and distance. These values will be returned as a tuple; tuples which are sequences similar to lists enclosed in parentheses per Section 1.6.3.

```
def projectile(v0, angle):
    """ Returns the projectile flight time, maximum height and ↙
        distance given initial velocity in meters per second and ↙
        launch angle in degrees. """
```

```python
from math import sin, cos
g = 9.8# gravity (meters per second squared)
angle_radians = 0.01745 * angle # convert degrees to radians
flight_time = 2 * v0* sin(angle_radians) / g
max_height = (1 / (2 * g)) * (v0 * sin(0.01745 * angle)) ** 2
distance = 2 * v0** 2 / g * sin(angle_radians) * cos(angle_radians)
return(flight_time, max_height, distance)
```

The `projectile` function is used next to compute the flight time, height, and distance for given inputs. Here the call is embedded within a print statement that prints the returned values.

```python
print(projectile(100, 45))
```

```
(14.4286123690312, 255.02644724730615, 1020.4081184645057)
```

Another way is to assign variables to each output for further processing or printing as below. The function returns them as a tuple of three, which is unpacked into three variables (see tuple Section 1.6.3). In subsequent examples, this is augmented to return lists containing time series of data for plotting the trajectories over time.

```python
v0 = 120
angle = 60
flight_time, max_height, distance = projectile(v0, angle)
print(f'A projectile with initial velocity {v0} m/s and angle ↙
    {angle} degrees will fly for {flight_time:.1f} seconds to a max ↙
    height of {max_height:.1f} and distance {distance:.1f} meters.')
```

```
A projectile with initial velocity 120 m/s and angle 60 degrees
will fly for 21.2 seconds to a max height of 550.9 and distance
1272.8 meters.
```

1.2.1   DEFAULT ARGUMENT VALUES

It is possible to specify default values for arguments enabling a function to be called with fewer arguments than it is defined to allow. They become optional function parameters with their default values set with an equals sign in the parameter list. This can save time and is flexible because the defaults can be overwritten when desired. For example, the parameter specification `print_flag=False` sets the default value to be used.

```python
def tank_volume(diameter, height, print_flag=False):
    PI = 3.14159
    volume = height * PI * (diameter/2) ** 2
    if print_flag == False:
        return(volume)
    else:
        print(f'Volume of a tank with diameter {diameter} and height ↙
            {height} is {volume(radius):.1f}')
```

The function can be called with just mandatory arguments or also include a subset of the optional parameters to override their defaults.

```
tank_volume(8, 20)      # mandatory arguments only
tank_volume(8, 20, True) # mandatory and optional arguments
```

The next method alternatively uses explicit keywords to define required positional parameters and/or optional parameters as function input.

1.2.2 KEYWORD ARGUMENTS

Functions can also be called using keyword arguments of the form `kwarg=value`. Keyword arguments must follow any provided positional arguments. All the keyword arguments passed must match one of the arguments accepted by the function, and their order is not important. This also includes non-optional arguments.

The previous volume function can be called several ways with or without using keywords:

```
tank_volume(12, 30)                       # 2 positional arguments
tank_volume(diameter=4, height=12)        # 2 keyword arguments
tank_volume(diameter=4, height=12, print_flag=True) # 3 keyword ↙
    arguments
tank_volume(print_flag=True, diameter=4, height=12) # 3 keyword ↙
    arguments different order
tank_volume(6, 15, True)                  # 3 positional arguments
tank_volume(6, 15, print_flag=True)       # 2 positional, 1 keyword
```

A keyword approach can make for more understandable code depending on the complexity of the function call (at the slight expense of verbosity), and it provides freedom in the ordering of sent parameters. This can be more robust compared to fixed ordering without keywords that can't always catch positional errors.

A *dictionary* data structure contains keyword pairs, as described in Section 1.10, which can be unpacked for function input with a double asterisk prefix such as `**kwargs`. The following unpacks an input dictionary equivalent to writing the inputs as `axis.set(xlabel='Time', ylabel='Height', ...)`.

```
axis_settings = {
    'xlabel': 'Time',
    'ylabel': 'Height',
    'title': 'Projectile Motion at given launch angles',
    }

axis.set(**axis_settings)
```

The above can be useful to set inputs once in a dictionary for sending to a function in many places. It then only needs to be modified in one place.

This function introduction established how code can be structured, modularized, and reused. Subsequent introductory topics, such as loops and conditionals, should be viewed with the mindset of creating reusable components. Knowing functions will also help in manipulating data structures. See Section 1.14 for more advanced function topics.

1.3 PRINT STATEMENTS

The print function has already been demonstrated with simple examples. The larger syntax details are introduced here and demonstrated later. The print function is used as a statement with the basic syntax:

```
print(arguments)
```

where the arguments are comma separated values, variables, or full expressions. They will all be converted to a string before printing to the output device, which defaults to the screen or optionally to a file. By default all arguments are separated by spaces in the output, but other separators can be specified in the full syntax.

```
weight = 450
print("The weight is", weight, "kilograms")

The weight is 450 kilograms
```

The full syntax for print is shown below and detailed in Table 1.1.

Print Syntax

```
print(arguments, sep=separator, end=end, file=file, flush=flush)
```

Table 1.1
Print Function Parameters

Parameter	Description
arguments	Arguments to print separated by commas.
sep	Optional. Character(s) used to separate the parameters, if there is more than one. Default is a space.
end	Optional. What to print at the output end. Default is a line feed (\n).
file	Optional. An object with a write method. Default is sys.stdout (normally the screen).
flush	Optional. A Boolean specifying if the output is flushed (True) or buffered (False). Default is False.

In the prior listing the variable weight is explicitly defined as an integer and prints cleanly. If the output is a floating point number as the result of computations, it may be desirable to format the output to limit the decimal places. Below shows a default print display containing undesirable insignificant digits. Formatting syntax to fix this is addressed in Section 1.9.

```
PI = 3.14159
radius = 5
volume = 4/3 * PI * radius**3
print('The volume of a sphere with radius', radius, 'is', volume)

The volume of a sphere with radius 5 is 523.5983333333332
```

1.4 OPERATORS

Operators are used to perform numerical and logical computations with given variables, data sequences, and values. The primary types are arithmetic, comparison, logical, identity, assignment, and membership operations. Bitwise operators are also available per the full reference at https://docs.python.org/3/reference.

1.4.1 ARITHMETIC OPERATORS

Arithmetic operators are used with numeric values to perform common mathematical operations per Table 1.2:

Table 1.2
Arithmetic Operators

Operator	Name	Example
+	Addition	x + y
–	Subtraction	x - y
*	Multiplication	x * y
/	Division	x / y
%	Modulus	x % y
**	Exponentiation	x ** y
//	Floor division	x // y

1.4.2 COMPARISON OPERATORS

Comparison operators in Table 1.3 are used to compare two values and to evaluate the result as True or False. These are typically used in Boolean expressions to check conditions for alternate logic paths. They can control conditional execution with if statements described in Section 1.5.1 and while loops in Section 1.8.2 and for loops in Section 1.8.1.

Table 1.3

Comparison Operators

Operator	Name	Example
==	Equal	x == y
!=	Not equal	x != y
>	Greater than	x > y
<	Less than	x < y
>=	Greater than or equal to	x >= y
<=	Less than or equal to	x <= y

```
>>> 5 ** 2 > 20
True
>>> 5 * 2 == 20
False
```

1.4.3 LOGICAL OPERATORS

Logical operators are used to combine conditional statements. These Boolean conditions are shown in Table 1.4. These are used in conjunction with comparison operators in Table 1.3 to perform more complex logical checks.

Table 1.4

Logical Operators

Operator	Description	Example
and	True if both statements are true	x < 5 and y < 10
or	True if one of the statements is true	x < 5 or y < 10
not	False if the result is true	not (x < 5 and y < 10)

1.4.4 ASSIGNMENT OPERATORS

Assignment operators are used to assign values to variables. The most common is the equals sign (=) but there are shortcuts available in Table 1.5 for concise representation of common operations that avoid repetitive patterns with variable names. The most frequent one is the increment operator += (also used to combine lists) that adds an amount to a variable by using x += 1 instead of the longer equation x = x + 1. It is frequently used in loops to update quantities.

Table 1.5
Assignment Operators

Operator	Description	Example	Equivalence
=	Simple assignment	x = 5	x = 5
+=	Add and assign[1]	time += dt	time = time + dt
-=	Subtract and assign	x -= 2	x = x - 2
*=	Multiply and assign	x *= multiplier	x = x * multiplier
/=	Divide and assign	x /= divisor	x = x / divisor
%=	Modulo and assign	x %= cycle_length	x = x % cycle_length
**=	Exponentiate and assign	x **= exponent	x = x ** exponent
//=	Floor divide and assign	x //= 3	x = x // 3

[1] also used to combine lists

Note that the = operator is a simple assignment, while the other operators perform the operation and then assign the result to the variable. The Table 1.5 equivalence column shows the equivalent operation using the = operator.

1.4.5   IDENTITY OPERATORS

Identity operators are used to compare objects, not to determine if they are equal values but if they are actually the same object residing in the same memory location. They are listed in Table 1.6. They may be used to compare different representations of the same data, help to debug code by checking if two objects are the same object, or optimize execution speed by avoiding unnecessary object creation.

Table 1.6
Identity Operators

Operator	Description	Example
is	True if both variables are the same object	x is y
is not	True if both variables are not the same object	x is not y

1.4.6   MEMBERSHIP OPERATORS

Membership operators are used to test if values are present or not in a sequence such as a list, tuple, set, or string. These can be useful for engineering data validation, filtering, and searching. They are described in Table 1.7.

Table 1.7

Membership Operators

Operator	Description	Example
in	True if a value is present in a sequence	x in list1
not in	True if a value is not present in a sequence	x not in list1

1.5   CONDITIONAL EXECUTION

Conditional execution of statement blocks is controlled with `if`, `elif`, and `else` statements in conjunction with logical, comparison, identity, and membership operators.

1.5.1   IF

The `if` keyword statement is used to specify a block of code to be executed if a logical condition is true. It precedes the block of code to be executed accordingly, and may be used in conjunction with `elif` and `else` statements expressed at the same level to delineate more conditional code blocks.

If Syntax

```
if condition:
  # block of code to be executed if the condition is true
```

An example to check a weight constraint is:

```
if weight > max_weight:
    print("weight is over the maximum")
```

1.5.2   ELSE AND ELIF

Use the `else` statement to specify a block of code to be executed when the `if` the condition is false.

If Else Syntax

```
if condition:
    # block of code to be executed if the condition is true
```

```
else:
    # block of code to be executed if the condition is false
```

With an `else` statement it prints results for both possible conditions:

```
# indentation of logical condition blocks
if weight > weight_max:
    print("Weight exceeds the maximum.")
else:
    print("Weight is within limits.")
```

The `else` can be used with a single `if` condition as above, or after other conditions using the `elif` statement described next. The `else` keyword will catch anything which isn't caught by preceding condition(s).

The `elif` keyword checks another condition when the previous condition is false. The following example uses it to check for three ranges of a variable.

```
if complexity > 10:
    risk = "High"
elif complexity > 5 and complexity <= 10:
    risk = "Medium"
else:
    risk = "Low"
```

Multiple `elif` statement blocks can be used under the top `if`. The last `else` statement isn't required with the other conditional checks and its use depends on the situational logic. Each condition is checked in order. If one is true, the corresponding block executes and no more conditions are checked. If more than one condition is true, only the first true one executes.

1.5.3   SHORT HAND CONDITIONAL STATEMENTS

SHORT HAND IF

If there is only one statement to execute, it can be put on the same line as the `if` statement to save space. A single line `if` statement is:

```
if weight > max_weight: print("Weight is too high.")
```

SHORT HAND IF ... ELSE

A single line can be used to express compound conditional statements with `if` and `else` by chaining the syntax for multiple conditions. This shortcut is called a *ternary* operator, or conditional expression, as a concise alternative to traditional if-else statements when choosing between values based on conditions.

Conditional Expression Syntax

```
some_expression if condition else other_expression
```

If there is one statement to execute for `if` and another one for `else`, it can all be put on the same line as below.

```
weight_ok = True if weight > max_weight else False
```

Multiple `else` statements can be put on the same line. This example replicates the multiple line logic of the three condition example in 1.5.2. One line contains the `if` and `else` statements for all three conditions. They are checked in the order shown.

```
> complexity = 3
> risk = "High" if complexity > 10 else "Medium" if complexity > 5 ↙
    else "Low"
> risk
'Low'
```

The previous example sets a value on the right side of an assignment statement (=), but functions or any evaluations can be performed as below.

```
print("Complexity is High") if complexity > 10 else ↙
    print("Complexity is Medium") if complexity > 5 else ↙
    print("Complexity is Low")
Complexity is Low
```

A contextual judgment should be made as to whether a ternary operator is more readable and appropriate than a multi-line conditional expression.

1.5.4   NESTED CONDITIONALS

Nesting of conditional `if` statements is possible with `if` statements contained inside other `if` statements.

```
if weight < weight_max:
    print("Met weight requirement")
    if camera_ready == True:
        print("Ready for flight")
    else:
        print("Camera is not ready")
```

1.5.5   AND

The `and` keyword is a logical operator that combines conditional statements and returns `True` if both statements are `True`.

```
if a > b and a < c: print("Both conditions are True")
```

1.5.6   OR

The or keyword similarly combines conditional statements and returns True if either statement is True.

```
if a > b or a < c: print("At least one of the conditions is True")
```

1.6   DATA STRUCTURES AND THEIR METHODS

Iterable data structures include lists, dictionaries, tuples, and sets. Each has its own methods or functions for performing operations on the data. These data structures are compared in Table 1.8 and detailed in following sections.

Lists are ordered collections appropriate for frequently modified data, very large datasets, and series of uniform data type, and as data containers within other lists or dictionaries. Dictionaries are unordered collections of key-value pairs suitable when there are logical associations between parameter keys and values, relationships between mixed data structures and types, complex structures, and networks. Tuples are best for data that cannot change after being assigned. They are sometimes preferable to lists for immutable data and execution efficiency. They can be useful as multidimensional keys in dictionaries. Sets are unordered collections of distinct non-duplicate items useful for analyzing membership of discrete items.

Table 1.8

Comparison of Lists, Dictionaries, Tuples, and Sets

Type	Notation	Element Access	Mutable
list	[23, 49, 18]	numeric index list_name[1]	Yes
dictionary	{'temp': 23, ↙ 'pressure': 49, ↙ 'humidity': 22}	key dict_name['pressure']	Yes
tuple	(23, 49, 18)	numeric index tuple_name[1]	No
set	{'gimbal', ↙ 'gps', 'frame'}[1]	None. Can loop through or check for item membership.	Yes

[1] Sets cannot have duplicate values, so a collection of measurements is generally not a feasible application.

1.6.1   LISTS

A list is a sequence of values which can be of any data type. The values in a list
are called elements or items. Lists are very useful as ordered collections of elements
which may include numbers, strings, other lists, dictionaries, tuples, and other data
types.

Lists are mutable because the elements can be changed. They are ideal for storing
and accessing data for which the number and value of elements are not known until
run-time, such as a simulation or real-time application.

There are many available built-in functions that return list information and methods that operate on lists including those in Table 1.9. The methods are accessed via
the dot notation, as shown in the examples for append and remove methods. The
functions take list names as inputs such as len. Following are examples of creating
and manipulating lists using some standard operations.

Table 1.9

List Methods and Functions

Method / Function	Description
append(item)	Adds an item to the end of the list.
extend(list)	Adds all the elements of a list to the end of the current list.
insert(index, item)	Inserts an item at the specified index.
remove(item)	Removes the first occurrence of an item.
pop()	Removes and returns the last item in the list.
pop(index)	Removes and returns the item at the specified index.
sort()	Sorts the items of the list in ascending order.
reverse()	Reverses the order of the items in the list.
index(item)	Returns the index of the first occurrence of an item.
count(item)	Returns the number of times an item appears in the list.
len(list)	Returns the length of a list.

DECLARING AND INITIALIZING LISTS

A list can be created using square brackets to enclose elements that are separated by
commas. The following creates two lists.

```
temperature_data = [60, 63, 55, 67]
planets = ["earth", 'mercury', 'mars']
```

The first example is a list of four integers assigned to temperature_data. The
second is a list of three strings assigned to planets. The elements of a list don't have

to be the same type. Lists can also be nested in other lists. The following list contains
a string, a float, an integer, and another list:

```
mixed_list = ['earth', 2.1 , 5, [15.1 , 23.5, 67.5]]
```

There are other ways to create a new list. An empty list contains no elements and can
be created with empty brackets: []. This would be necessary to initialize a list prior
to a loop where data values are appended to it.

ACCESSING ELEMENTS WITHIN A LIST

Elements within a list can be accessed using numeric indices. The index of an element in a list starts from 0, so to access the first element in the list use the following:

```
first_temperature = temperature_data[0]
print(first_temperature)
60
```

Negative indices can also be used which start from the end of the list, with -1 being
the last element, -2 the second to last, etc. The same notation is used for slicing of
list elements by ranges and steps overviewed in Section 1.13.

UPDATING LIST ELEMENTS

Elements within a list can be updated by assigning new values to them using the
indexing syntax. To change the first temperature value in the list refer to it through
its index:

```
temperatures[0] = 295
```

APPENDING ELEMENTS TO A LIST

Elements can be added to the end of a list using the append method.

```
temperature_data.append(64)
print(temperature_data)
[60, 63, 55, 67, 64]
```

Lists can be generated from other lists. inputs Consider the problem of calculating the internal energy of a gas at different temperatures using the equation
$U = n * R * T$, where n is the number of moles of gas, R is the ideal gas constant, and
T is the temperature. Lists of temperatures and their corresponding internal energies
can be created for analysis and plotting as below.

```
temperatures = [ 293, 300, 310, 330 ]
# initialize data list
internal_energies = []
n = 1 # number of moles of gas
```

```
R = 8.31 # ideal gas constant

for temperature in temperatures:
    U = n * R * temperature
    # append value to list
    internal_energies.append(U)

print(internal_energies)
```

```
[2380.23, 2427, 2503, 2673]
```

Alternatively, one can use plus sign syntax (+) to add lists of one or more elements each:

```
temperature_data = temperature_data + [current_temperature]
```

REMOVING ELEMENTS FROM A LIST

Elements can be removed from a list using the `remove` method. To remove a specific temperature value, use the following.

```
temperature_data.remove(67)
```

Another way to remove specific elements is to redefine a list by slicing it with numeric indices. The following new list starts at the second value indicated with index 1 before the colon, and the blank after it indicates through the last element. This removes the first value in the list.

```
# remove initial value by slicing from the second element to the end
temperature_data = temperature_data[1:]
```

CONCATENATING LISTS

The + operator concatenates separate lists.

```
inner_planets = ["earth", 'mercury', 'mars', 'venus']
outer_planets = ["Jupiter", "Saturn", "Uranus", "Neptune", "Pluto"]
solar_system_planets = inner_planets + outer_planets
print(solar_system_planets)
```

```
["earth", 'mercury', 'mars', 'venus', "Jupiter", "Saturn",
"Uranus", "Neptune",  "Pluto"]
```

```
test1_output = [11, 12, 23]
test2_output = [24, 15, 16]
total_output = test1_output + test2_output
print(total_output)
```

```
[11, 12, 23, 24, 15, 16]
```

SORTING

The sort method arranges the elements of the list from low to high:

```
>>> temperature_data.sort()
>>> temperature_data
[55, 60, 63, 64, 67]
```

REPEATING A LIST

The * operator repeats a list a given number of times. It can be useful for initializing list values.

```
>>> num_values = 4
>>> [0] * num_values
[0, 0, 0, 0]
>>> [1, 2, 3] * 3
[1, 2, 3, 1, 2, 3, 1, 2, 3]
```

ITERATING OVER A LIST

Lists can be easily iterated over. The most common way to traverse the elements of a list is with a standard for loop:

```
for temperature in temperatures:
    print(temperature)
```
```
295
310
330
340
```

In order to write or update lists elements, the indices are needed. A common way is to combine the functions range and len. The loop below iterates through the list elements and updates each one. The range function returns an iterable list of indices from 0 to $n-1$, where n is the length of the list from the len function. Each time through the loop i gets the index of the next element, and the assignment statement uses i to read the element value and assign the new value.

```
# apply calibration factor to dataset
for i in range(len(dataset)):
    dataset[i] = dataset[i] * 2.94
```

Note a more elegant and concise way is to use a *list comprehension* and redefine the list as follows. See Section 1.6.5 for more details on comprehensions.

```
dataset = [datapoint * 2 for datapoint in dataset]
```

1.6.2   DICTIONARIES

Python dictionaries are data structures that store key-value pairs. They are similar to lists, but instead of having a numerical index, dictionaries have keys that can be any immutable type such as strings or numbers or tuples. The values can be of any data type including lists or additional dictionaries at any level of nesting. This makes dictionaries a powerful data structure for storing and accessing data in a logical, organized manner. They are essential for many applications.

A dictionary provides a mapping where each key maps to a value. Each key is unique and can be used to access the values as indices. The association of a key and a value is called a key-value pair or an item. When viewing a dictionary, the items are displayed in a certain order, but it is irrelevant. Data is extracted the same way, hence they are called unordered vs. ordered lists.

These features make dictionaries more flexible than lists. They don't displace the usage of lists because lists can be the associated values in a dictionary.

Table 1.10 summarizes the available dictionary methods and functions to perform operations on them.

CREATING A DICTIONARY

The syntax for a dictionary is enclosing a comma-separated list of key-value pairs in curly braces as {}. The key-value pairs are separated by colons, and each key-value pair is separated by a comma.

The mass characteristics for the components of a small satellite can be populated as below.

```
satellite_part_masses = {'frame': 310, 'gimbal':491, 'gps antennae': ↙
    311, 'gps receiver': 480, 'stellar gyroscope': 52}
```

Dictionary values are flexible to hold entire data structures including lists, dictionaries, or other objects. The following is a data capture of serial measurements against time from individual sensors. The keys are the measured quantities and values are lists of the measurements.

```
sensor_data = {'temperature': [62.5, 62.1 , 64.4, 63.7, 60.0, 61.1 ↙
    ], 'pressure': [12.4, 13.9, 14.8, 12.5, 15.6, 17.1 ]}
```

ACCESSING DICTIONARY VALUES

Values in a dictionary can be accessed using square braces [] containing a specific key name.

```
temperature = sensor_datapoint['temperature']
```

```
print('gimbal mass is', satellite_part_masses['gimbal'], 'grams')
gimbal mass is 491 grams
```

Table 1.10
Dictionary Methods and Functions

Method/Function	Description
`dict()`	Creates a new dictionary
`get(key, default=None)`	Returns the value of the specified key, default if key doesn't exist
`keys()`	Returns a view object containing the keys of the dictionary
`values()`	Returns a view object containing the values of the dictionary
`items()`	Returns a view object containing the key-value pairs of the dictionary
`pop(key, default=None)`	Removes and returns the value of the specified key, default if key doesn't exist
`popitem()`	Removes and returns an arbitrary key-value pair from the dictionary
`clear()`	Removes all key-value pairs from the dictionary
`copy()`	Returns a shallow copy of the dictionary
`update(other_dict)`	Updates the dictionary with key-value pairs from other dictionary
`fromkeys(seq, value=None)`	Creates a new dictionary from a sequence with specified value
`len(dict)`	Returns the number of key-value pairs in the dictionary
`key in dict`	Returns True if the key exists in the dictionary, False otherwise
`dict[key] = value`	Assigns a value to the specified key
`del dict[key]`	Removes the key-value pair from the dictionary

ADDING AND UPDATING DICTIONARY ITEMS

New dictionary items can be added with an assignment statement using square brackets and the new key name.

```
satellite_part_masses['battery'] = 834
satellite_part_masses['thermal_camera'] = 56
```

If a key already exists in the dictionary, update its value by reassigning a new value to the key:

```
satellite_part_masses['battery'] = 764
```

REMOVING DICTIONARY KEY-VALUE PAIRS

To remove a key-value pair from a dictionary, you can use the `del` keyword and the key name:

```python
del satellite_part_masses['thermal_camera']
```

EXTRACTING DICTIONARY ITEMS

The methods for extracting data from a dictionary by iterating over them are:

- `.keys()` returns the dictionary keys as a list
- `.values()` returns the dictionary values as a list.
- `.items()` returns both keys and values as a list of (key, value) tuples.

The total satellite mass can be calculated by iterating over the dictionary values with the `.values()` method.

```python
satellite_mass = 0
for mass in satellite_part_masses.values():
    satellite_mass += mass
satellite_mass
1644
```

A condensed version would be the sum of a list comprehension .

```python
sum([mass for mass in satellite_part_masses.values()])
```

The `.items()` method returns both keys and values.

```python
for part, mass in satellite_part_masses.items():
  print(part, '->', mass)
frame -> 310
gimbal -> 491
gps antennae -> 311
gps receiver -> 480
stellar gyroscope -> 52
```

Dictionaries are very convenient in many situations. Keyword arguments can be used in function calls per Section 1.2. They provide complete flexibility in the order of function arguments, and the data being passed is explicitly named which reduces errors.

Below a dictionary of graph options is sent to Matplotlib using the ** syntax for *keyword arguments* to unpack the dictionary. This way a variable number of arguments can be passed using ** as a wildcard notation. Such a dictionary can be defined in one place and referenced in many function calls as a form of reuse and to enforce consistency. It also eliminates writing the same arguments in many places. This method is illustrated in Section 2.2 with Matplotlib and subsequent examples.

```
# Dictionary for default histogram keyword parameters
kwargs = {'color': "blue", 'rwidth': 0.9}

figure, ((axis1, axis2), (axis3, axis4)) = plt.subplots(2, 2)
...
axis1.hist (station1, num_bins, **kwargs)
...
axis2.hist (station2, num_bins, **kwargs)
...
axis3.hist (station3, num_bins, **kwargs)
...
```

NESTED DICTIONARIES

Dictionaries can contain other dictionaries nested at any level, which is useful in engineering applications containing data and objects representing systems in a natural way. They serve as multi-level relational databases. They are used internally in major Python packages as primary data structures due to their versatility. The often used term *dictionary of dictionaries* superficially implies two levels, but there can be many levels of dictionaries.

For example, the satellite parts dictionary can be enhanced for a more complex satellite design to represent and analyze it at different design levels with a nested dictionary. This allows mass to be expressed at any hierarchical level appropriate to the architecture.

```
satellite_part_masses = {'frame': 310,
                    'gimbal':491,
                    'gps': {'gps antennae': 311, 'gps receiver': 4↙
                        80},
                    'stellar gyroscope': 52}
```

A computable engineering project plan is represented in the following nested dictionary of four levels. The top keys are the project names, and the values are sub-dictionaries. Contained within those project dictionaries are additional key-value pairs where some of the values are dictionaries themselves. Here the lowest level dictionary items correspond to each task.

```
projects = {
    'Laser Calibration': {
        'start_date': '2022-04-01',
        'end_date': '2022-06-30',
        'tasks': {
            'setup': {
                'duration': 14,
                'assigned_to': 'Jose Omega'
            },
            'data analysis': {
```

```
            'duration': 11,
            'assigned_to': 'Ada Knuth'
        }
    }
},
'Board Development': {
    'start_date': '2022-07-01',
    'end_date': '2022-09-30',
    'tasks': {
        'design': {
            'duration': 23,
            'assigned_to': 'Data Pascal'
        },
        'fabrication': {
            'duration': 18,
            'assigned_to': 'Rosetta Stone'
        }
    }
}
}
}
```

The following example will use the project dictionary to compute the total number of person-days in the collective tasks. It sums the durations from the individual task dictionaries where each value is referenced with the `duration` key. This dictionary structure will be revisited in Chapter 2 network analysis and graphing applications.

```
total_person_days = 0

for project in projects.values():
    for task in project['tasks'].values():
        duration = task['duration']
        total_person_days += duration

print(f"Total Multi-Project Effort = {total_person_days} Person Days")
Total Multi-Project Effort = 66 Person Days
```

This demonstrates how a dictionary may serve as a relational database. A spreadsheet could not be used for the same due to the information nesting of multiple tables.

1.6.3 TUPLES

Tuples are similar to lists, being sequences of values that can be of different data types. The primary difference is that tuples are immutable because their values cannot be changed once they are created, as opposed to lists where items can be changed, deleted, or added. This makes tuples useful for representing data that should not be modified. A summary of tuple methods and functions is in Table 1.11.

Table 1.11
Tuple Methods and Functions

Method / Function	Description
`count(x)`	Returns the number of times x appears in the tuple.
`index(x)`	Returns the index of the first occurrence of x in the tuple. Raises a `ValueError` if x is not found.
`len()`	Returns the length of the tuple.
`sorted()`	Returns a sorted list of the elements in the tuple.
`max()`	Returns the largest element in the tuple. Raises a `TypeError` if the elements are not comparable.
`min()`	Returns the smallest element in the tuple. Raises a `TypeError` if the elements are not comparable.
`tuple(iterable)`	Converts an iterable (e.g., a list or a string) into a tuple.

CREATING A TUPLE

Tuples are enclosed in parentheses as `(x_coord, y_coord, z_coord)`. A tuple of compounds is:

```python
compounds = ('Silicon Dioxide', 'Boron Nitride', 'Gallium Arsenide')
```

ACCESSING TUPLE VALUES

Similar to lists, use the index notation to identify particular elements in a tuple.

```python
print(compounds[0])
```
```
Silicon Dioxide
```

Due to the restrictions of immutability, tuples may seem less capable than lists but they are computationally more efficient. They use less memory and can be operated on faster. They are often used as the return type from functions that return more than one value. There is also a benefit to limiting possible errors by restricting data structure updates, which becomes more important with large and complex software.

In general it is a good idea to use a tuple instead of a list, unless a list is the only way to process data, When necessary a list can always be created from a tuple, or vice-versa.

```python
list(compounds)
```
```
['Silicon Dioxide', 'Boron Nitride', 'Gallium Arsenide']
```

1.6.4 SETS

A set is an unordered collection of unique elements. Sets are useful for storing items where duplicates are not allowed and when the order of elements is not important. Sets can be used to perform mathematical operations for union, intersection, difference, and symmetric difference. They are mutable, since elements can be added or removed after set creation. Set methods and functions are shown in Table 1.12.

Sets are efficient for membership tests, allowing one to check whether an element is in the set quickly. They are useful to perform operations on data collections where the sequence or frequency of elements is irrelevant, such as when identifying unique values from a dataset or comparing different groups.

Table 1.12

Set Methods and Functions

Method / Function	Description
`add(x)`	Adds the element x to the set. If x is already present, no change occurs.
`remove(x)`	Removes the element x from the set. Raises a `KeyError` if x is not found.
`discard(x)`	Removes the element x from the set if it is present. No error is raised if x is not found.
`union(other)`	Returns a new set containing all elements from the set and `other`. Equivalent to | operator.
`intersection(other)`	Returns a new set containing elements common to the set and `other`. Equivalent to & operator.
`difference(other)`	Returns a new set containing elements in the set but not in `other`. Equivalent to - operator.
`symmetric_difference(other)`	Returns a new set containing elements in either set or `other` but not both. Equivalent to ^ operator.
`len()`	Returns the number of elements in the set.
`clear()`	Removes all elements from the set.
`set(iterable)`	Converts an iterable (e.g., a list) into a set.

A set is defined using curly braces {} or the `set()` function. In the next example, `uav_components` is a set with four elements, and `unique_numbers` will contain only the unique values {1, 2, 3, 4}, as duplicates are automatically removed.

```python
# Defining sets
uav_components = {'GPS', 'Camera', 'Battery', 'Propeller'}
unique_numbers = set([1, 2, 2, 3, 4])
```

Python provides several methods for working with sets, such as adding elements, removing elements, or getting the number of elements:

```python
uav_components.add('IMU')  # Adds an element
uav_components.remove('Battery') # Removes an element
print(uav_components)
print(len(uav_components))

{'Propeller', 'GPS', 'IMU', 'Camera'}
4
```

Sets also support mathematical operations:

```python
# Union
all_equipment = uav_components.union({'Parachute', 'Antenna'})
# Intersection
common_parts = uav_components.intersection({'Camera', 'Propeller'})
# Difference
missing_parts = uav_components.difference({'Battery', 'Antenna'})
# Symmetric Difference
diff = uav_components.symmetric_difference({'Camera', 'GPS'})

print('union:', all_equipment)
print('intersection', common_parts)
print('difference', missing_parts)
print('symmetric difference:', diff)

union: {'Propeller', 'Antenna', 'Parachute', 'GPS', 'IMU', 'Camera'}
intersection {'Propeller', 'Camera'}
difference {'Propeller', 'GPS', 'IMU', 'Camera'}
symmetric difference: {'Propeller', 'IMU'}
```

Sets are efficient for membership tests due to their underlying implementation as hash tables, making them ideal for checking if an element is part of a collection quickly:

```python
if 'GPS' in uav_components:
    print('GPS is installed on the UAV.')

GPS is installed on the UAV.
```

1.6.5 DATA STRUCTURE COMPREHENSIONS

A *comprehension* has been introduced as a shortcut method for creating and populating sequences for lists, dictionaries, and sets via a set of looping and filtering instructions. It consists of a single expression followed by at least one `for` clause and optionally includes one or more `if` clauses.

A comprehension is contrasted with a standard way to populate a list using a multi-line `for` loop as below. This example creates a list of random values that will be reduced to a single line in a *list comprehension*.

```
import random
# initialize list of windspeeds
random_windspeeds = []

# populate the list of random windspeed values over time uniformly ↙
    distributed between 0 and 30
for hour in range(24):
    random_windspeeds.append(random.uniform(0, 30))
```

The above can be expressed in a single line to initialize and populate the list by enclosing the comprehension in brackets:

```
random_windspeeds = [random.uniform(0, 30) for hour in range(24)]
```

The same comprehension below in the interpreter will print all the values from the single statement. The interpreter evaluates the expression and prints it per a REPL loop.

```
>>> [random.uniform(0, 30) for hour in range(24)]
[10.714446850428274,
 7.39771612646417,
 20.270621799337654,
...
```

The list comprehension could also be provided as a `print` function input parameter that is evaluated to display the values as below. This is an example of nested expressions and functions, e.g. the comprehension expression evaluation is directly input to the print function.

```
print([random.random() for hour in range(24)])
```

List comprehensions can combine multiple `for` loops nested logic as below. During execution it sets the outer index for tanks to its initial value, iterates through the inner loop for volume quantities, then goes back to the outer loop for the next tank value and continues.

```
>>> [tanks * volume for tanks in [1, 2, 3, 4] for volume in [100, 200↙
    , 300]]
[100, 200, 300, 200, 400, 600, 300, 600, 900, 400, 800, 1200]
```

A dictionary comprehension needs to set the keys and values. The following creates a dictionary from two lists for compound data.

```
compounds = ["Silicon Dioxide", "Boron Nitride", "Gallium Arsenide"]
molecular_weights = [60.088, 24.813, 144.645]

# Dictionary comprehension for molecular weights
compound_weights = {compounds[i]: molecular_weights[i] for i in ↙
    range(len(compounds))}

print(compound_weights)
```

```
{'Silicon Dioxide': 60.08, 'Boron Nitride': 24.81, 'Gallium Arsenide': 144.64}
```

Next a larger dictionary of compound data is filtered to create a new reduced dictionary. Here the list of available compounds is screened for a constraint on the melting temperature with an `if` statement in the expression.

```
compound_data = {
    "Silicon Dioxide": {
        "symbol": "SiO2",
        "molecular_weight": 60.088,
        "melting_point": 1610, # °C
        "thermal_conductivity": 1.4, # W/(m*K) (average value)
    },
    "Boron Nitride": {
        "symbol": "BN",
        "molecular_weight": 24.813,
        "melting_point": 3000, # °C (approximate)
        "thermal_conductivity": 27.6, # W/(m*K)
    },
    "Gallium Arsenide": {
        "symbol": "GaAs",
        "molecular_weight": 144.645,
        "melting_point": 1238, # °C
        "thermal_conductivity": 50, # W/(m*K) (approximate)
    },
}

# Create dictionary of compounds with melting points less than 1800
melting_points = {compound: compound_data["melting_point"] for ↙
    compound, compound_data in compounds.items() if ↙
    compound_data["melting_point"] < 1800}

print(melting_points)
{'Silicon Dioxide': 1610, 'Gallium Arsenide': 1238}
```

The subject of a comprehension is very general and can be used for more than assigning values. The leading expression will be evaluated if there is no equivalence statement using an equals sign. Below illustrates how expressions and functions can be nested. Legibility and maintenance should always be considered before nesting too many levels however. Here for example, the computation may be better suited outside of the `print` statement in the midst of a larger program.

```
[print (velocity, velocity**2/g/2) for velocity in range(0, 101, 10)]
0 0.0
10 5.1020408163265305
20 20.408163265306122
30 45.91836734693877
...
```

1.7   SLICING

Slicing is a way to extract a subset of elements from data types such as lists, tuples, and strings. It allows one to create new sequences by selecting specific elements from the original sequence based on their indices. It uses the following syntax below with the parameters in Table 1.13.

Slicing Syntax

```
sequence[start:stop:step]
```

Table 1.13
Slicing Parameters

Parameter	Description
start	The index at which the slicing begins (inclusive). If omitted, it defaults to 0.
stop	Optional. The index at which the slicing ends (not inclusive). If omitted, it defaults to the length of the sequence.
step	Optional. The step size between elements. If omitted, it defaults to 1. It can be negative

```python
parts = ["controller", "frame", "battery", "wire harness", "GPS"]

# remove last element
print(parts[:-1])

# slice from index 1 to 3 (not inclusive)
print(parts[1:4])

# slice until the end
print(parts[2:])
['controller', 'frame', 'battery', 'wire harness']
['frame', 'battery', 'wire harness']
['battery', 'wire harness', 'GPS']
```

Slicing is a powerful and concise way to manipulate sequences in Python by extracting specific portions of a sequence or create new sequences based on requirements. Slicing can be particularly useful when working with large datasets or when performing operations on subsets of data.

1.8 ITERATION AND ITERABLES

Iteration is the repetition (looping) of program statements based on sets of elements to process or with conditional logic to stop the repetition. An iterable is an object capable of returning its members one at a time for iteration. Iterables can be used in a `for` loop and many other places where a sequence is needed such as input to the `zip` and `map` functions, and other operations.

Iterables include all sequence types such as lists, tuples, or strings and some non-sequence types like dictionaries, file objects, and objects of any classes defined with a `__iter__()` method or with a `__getitem__()` method. The latter methods are not always required to use with iterables but are available.

The data structures described earlier are iterables. Table 1.14 summarizes the methods. Tuples do not have comprehensions , as the parentheses create a generator instead.

Table 1.14

Iteration Methods for Lists, Dictionaries, and Tuples

Data Structure / Iteration Method	Example
Lists	
`for` loop	`for item in list_name:`
	`for temperature in temperature_list:`
`for` loop with `enumerate()`	`for datapoint, temperature in` ↙ `enumerate(temperature_list):`
list comprehension	`[item for item in list_name]`
Dictionaries	
`for` loop over keys	`for key in dict_name:` [1]
	`for chemical in chemical_data.keys():`
`for` loop over values	`for value in dict_name.values():`
	`for molecular_weight in` ↙ `chemical_data.values():`
`for` loop over key-value pairs	`for key, value in dict_name.items():`
	`for chemical, molecular_weight in` ↙ `chemical_data.items():`
dictionary comprehension	`{key:value for key, value in` ↙ `dict_name.items()}`
Tuples	
`for` loop	`for item in tuple_name:`

[1] `.keys()` after the dictionary name is optional.

1.8.1   FOR LOOP

The `for` statement introduced in Chapter 0 is used to iterate over the elements of any iterable sequence including lists, dictionaries, tuples, strings, and other iterable objects. The nested statements will be repeated as a block for each item in the sequence where the `variable` name will take on the value of each item.

For Loop Syntax

```python
for variable in sequence:
    # block of code to be executed for each element in sequence
```

The `range()` function is often used in a `for` statement to loop through the code block a specified number of times. It is specified as `range(start, stop, step)` and returns a sequence of integers from `start` up to, but not including the `stop`. The `start` parameter is optional and defaults to zero and increments with an optional `step` size of 1 by default.

The following will print the hours 1 through 24, each on a separate line. The more powerful NumPy `linspace` function is frequently used instead of `range` and is not limited to integer values. See examples in later chapters.

```python
for hour in range(1, 25):
    print(hour)
```

Another way to control a `loop` is to loop is to iterate over a list.

```python
for part in ["controller", "frame", "battery", "wire harness"]:
    print(part)
```

1.8.2   WHILE LOOP

The `while` control statement is used for repeated execution of a statement block while an expression is true. The "while loop" syntax is below where the indented code block of any arbitrary size is repeated.

While Loop Syntax

```python
while condition:
    # block of code to be executed if the condition is true
```

An example is a loop for computing dynamic variables over time up to a fixed end time. The logical condition `time < end_time` would end the loop appropriately when it evaluates to `false`.

```python
time = 0
end_time = 24
timestep = .1
while time < end_time:
```

```
# block of code to be executed each time step
time += timestep
```

When deciding between `while` and `for` loops, the choice depends on the nature of the iteration and the control needed over the loop's execution. A `for` loop is suited for situations where the number of iterations is known in advance, such as iterating over elements in a list, range, or other iterable where the number of repetitions is fixed. In contrast, a `while` loop is appropriate when the number of iterations is not predetermined and is instead dependent on a condition being met. This is useful when execution must continue until a certain logical condition changes, such as in a random simulation or reading data until the end of a file.

1.8.3   ENUMERATE LOOP

Sometimes a numeric index is needed during iteration. The `enumerate` function is similar to a `for` loop but also returns the updated index with the item for each iteration (as a tuple). In the next example it is desired to place n system processes in a grid for a sequence diagram (using graphviz). To visually arrange the objects and actions, x coordinate values are calculated using the index from the enumerate function for `actor_object_number` and spacing accordingly.

```
object_spacing = 3
object_x_coords = {} # dictionary holds x coordinates for action ⤸
    nodes and edges
for actor_object_number, actor_object in enumerate(actors+objects, ⤸
    start=1):
    x = int(actor_object_number-1)*object_spacing
    if actor_object in actors:
        g.node(actor_object, pos=f"{x},.2!") # graphviz node
    object_x_coords[actor_object] = x
```

1.9   NUMBER AND STRING FORMATTING

Engineers need to convey data and results in a clear and efficient format, both for human interpretation and for further processing. The output data may be file input for additional analysis or serve as AI training data for machine learning. Significant digits should be considered while suffering no accuracy loss. This section provides a distillation of numerous formatting options with some recommendations.

Formatting of numeric and string output can be specified three ways as summarized in Table 1.15. F-strings are highly recommended and used throughout this book. The string `format` method is more limited and harder to use. The % string formatting is not recommended but retained in legacy code and some generated solutions, so one should be aware of it.

Table 1.15 shows equivalent examples to cleanup the raw result in Section 1.3 for the calculation `volume = 4/3 * PI * radius**3` given `radius = 5`. They all produce the same output.

Each method in Table 1.15 prints `radius` as a default string (in this case it's a clean integer) and `volume` as a fixed point number with a single digit of precision denoted by `.1 f`. The f-string method examples show that single variables or full expressions can be contained in a string for evaluation.

Table 1.15

Formatting Methods Summary and Equivalent Examples

All examples produce the same output:
`Volume of a sphere with radius 5 is 523.6`

Method	**Summary and Equivalent Examples**
f-strings	Recommended for all new development. The most intuitive, flexible and powerful formatting technique. Expressions are embedded in a string with curly brackets `{...}` prefixed with `f` or `F` with an optional format specifier after the expression. The variables and expressions inside the brackets form strings to be evaluated at runtime.

```
print(f'Volume of a sphere with radius {radius} is ↵
    {volume:.1f}')
```

```
print(f'Volume of a sphere with radius {radius} is ↵
    {4/3 * math.pi * radius **3:.1f}')
```

string `format` method	A string method with superior capabilities to % string formatting but superseded by f-strings. Requires each printed parameter to be lined up and accounted for in two places within the string using brackets before the `format` method and in its input expressions.

```
print('Volume of a sphere with radius {} is ↵
    {:.1f}'.format(radius, volume))
```

% string formatting	Avoid using because it is difficult to write and has more primitive formatting capabilities. It will go away but may be in old legacy code and generated solutions. Difficult to use as each printed parameter needs to be accounted for in two places separated by the % symbol, first in the string itself using a % symbol and in a tuple after the separating % symbol.

```
print('Volume of a sphere with radius %s is %.1f' % ↵
    (radius, volume))
```

1.9.1   F-STRINGS

F-strings are highly recommended due to their flexibility and convenience for more legible coding, and to produce better output. They are more readable, concise, have more options, are less prone to error, and execute faster than other formatting methods.

F-strings are prefixed with `f` or `F` and allow the inclusion of expressions inside the string with curly brackets `{expression}` with optional format specifiers after the expression following a colon `:`. The full curly bracket syntax for an f-string with all options is below where the square brackets denote options:

F-String Expression Syntax

`{expression[=]:[flags][width][.precision][type]}`

The *flags* are optional parameters for text alignment and number notations. The parameters and examples are shown in Tables 1.16 through 1.19. The *width* parameter is used to denote the total space to write an output element, the *precision* after the decimal point refers to how many numbers to display after the decimal point (valid for floating point numbers), and *type* identifies the expression format type (e.g., floating point options, integers, etc.).

String is the default formatting type and does not need to be specified. If one needs to print a simple string variable, then no modifiers or data type specification are necessary. The variable or fuller expression will be printed starting at the left bracket position within the f-string.

```
engine_type = "turbocharged"
print(f"Engine type: {engine_type}")
```

`Engine type: turbocharged`

Dealing with floating point numbers is not always as simple. An illustration of the f-string options for a fixed point number are in Figure 1.1. The formatting syntax `6.2f` denotes a total width of 6 spaces with 2 digits of precision after the decimal point. Other examples for this `6.2f` format are in Table 1.16.

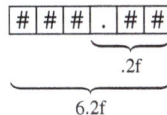

Figure 1.1   Example Fixed Point Number Spacing for Width = 6, Precision = 2

The `.2f` format specifier alone will print a floating point number rounded off to two decimal points. The field width can also be specified to align the output with the full `:6.2f` to print a floating point number six spaces wide.

Table 1.16

Float Number Character Formatting Examples for 6.2f

Value	Display					
723.566	1	2	3	.	5	6
472.7	4	7	2	.	7	0
0.029			0	.	0	3
13		1	3	.	0	0
9423.17	9	4	2	3	.	1

```python
print(f'Kinetic Energy = {kinetic_energy:6.2f} Joules')
```

```
Kinetic Energy = 371.31 Joules
```

Shortcut f-string Syntax to Print Expression

An optional equals sign = can be added after a variable or larger expression, and it will print both the expression and its value as a shortcut. E.g., instead of the previous example one can write the following and it will print the variable name with an equals sign before its value. Any number of spaces can be added around the equals sign while using the usual format specifiers.

```python
print(f"{kinetic_energy = :6.2f}")
```

```
kinetic_energy = 371.31
```

More complex expressions can be contained and will be included in the output string as written.

```python
print(f"{4/3 * PI * radius ** 3 = :.1f}")
```

```
4/3 * PI * radius ** 3 = 523.6
```

This shortcut can be useful when working interactively in the interpreter command line mode or adding print statements to debug a script. The shortcut f-string expression can also be used conveniently without a print statement. It can serve as temporary scaffolding to print expressions or variables throughout the code in development and debugging.

1.9.2 STRING FORMAT METHOD

A string is a Python class object with a `format` method that is invoked with the dot prefix notation. The basic syntax for the `format` method is below.

String Format Method Syntax

```python
string.format(expression(s))
```

The string contains replacement fields enclosed in curly brackets {...} for all the expression(s) contained in the call separated by commas (a tuple). The expressions are either positional or keyword arguments.

Table 1.15 shows an example using positional arguments for the `radius` and `volume` variables. The same result can be obtained with keyword arguments per the following.

```
print('Volume of a sphere with radius {radius} is ⤸
    {volume:.1f}'.format(radius = 5, volume = 4/3 * pi * r**3))

Volume of a sphere with radius 5 is 523.6
```

Written code using `.format()` is more readable than using % string formatting described next, but can still be quite verbose when dealing with multiple parameters and longer strings. All expressions must be accounted for in two places vs. one with an f-string. The added complexity also leads to more errors.

1.9.3 % STRING FORMAT

This more primitive format uses the *string modulo operator* % where on the left side is the string to format and on the right side is a tuple with the content to insert. The content is interpolated into the format string similar to the `format` method. The fields and expressions must line up by position on the left and right sides of the modulo operator.

```
print("Volume for sphere with radius %2.2f = %2.2f" % (r, volume))

Volume of a sphere with radius 5.00 is 523.6
```

When using several parameters and longer strings, the code will quickly become less readable and messy.

1.9.4 FORMATTING OPTIONS

Many formatting options are available for numeric display styles and alignment. The options in Table 1.17 for floating point numbers and Table 1.18 for integers apply to f-strings and the string `.format()` method. A subset will work with % string formatting.

Basic number alignment options are in Table 1.19. Examples for the alignments in a field of six characters are in Table 1.20. An example left alignment without specifying a field width is:

```
for velocity in range(0, 50, 20):
  print(f"{velocity:<} {(velocity**2):<}")
0 0
20 400
40 1600
```

Table 1.17

Float and Decimal Number Format Options

Given: `c = 299792458 #m/s, pi = 3.141592653589793`

Option	Description	Examples
:f	Fixed point with decimal (default precision 6)	`print(f'{pi = :f}')` `pi = 3.141593`

:.nf	Fixed point with *n* digits of precision after decimal point.	`print(f'{c = :.0f} m/s \n{pi = :.2f}')`
		`c = 299792458 m/s`
		`pi = 3.14`

:e	Scientific format with lower case e	`print(f'{c = :e} m/s \n{pi = :.3e}')`
		`c = 2.997920e+08 m/s`
		`pi = 3.142e+00`

:E	Scientific format with upper case E	`print(f'{c = :E} m/s \n{pi = :.3E}')`
		`c = 2.997920E+08 m/s`
		`pi = 3.142E+00`

:g	General format	`print(f'{c = :g} m/s \n{pi = :.3g}')`
		`c = 2.99792e+08 m/s`
		`pi = 3.14`

:G	General format using a upper case E for scientific notations	`print(f'{c = :G} m/s \n{p = :G}')`
		`c = 2.99792E+08 m/s`
		`pi = 3.14159`

Table 1.18 shows format options for integer numbers. There are additional format notations available, which can be adjusted based on precision needs. Case sensitivity affects how special values are displayed, and formatting can adapt between fixed-point and scientific styles depending on the number's size. Locale-aware formatting ensures numbers are displayed according to regional conventions. These more detailed formatting options for float, decimal numbers, and strings are at https://docs.python.org/3/library/string.html including unicode conversion, hex format, and others.

Table 1.18

Integer Number Format Options

Option	Description	Example
:d	Decimal format	`print(f'{c = :d} m/s')` `c = 299792458 m/s`
:b	Binary format	`print(f'255 integer = {255:b} binary')` `255 integer = 11111111 binary`
:o	Octal format	`print(f'255 octal = {255:o}')` `255 octal = 377`

Table 1.19

Number Alignment Options

Description	Option
Left align	:<
Left align with a field width of n	:<n
Center within field	:^
Center align with a field width of n	:^n
Right-align	:>
Right-align with a field width of n	:>n
Right align with a field width of n, fill with 0s	:0>n
Use space instead of sign for positive numbers	:

The next example left justifies output for a fixed field width including a header row.

```
# 8 character columns left justified with '<'
print (f"{'velocity':<8} {'height':<8}")
print(f"{velocity:<8.2f} {height(velocity):<8.2f}" for velocity in ↙
    range(0, 50, 10))

velocity height
0.00     0.00
10.00    1.56
20.00    6.25
30.00    14.06
40.00    25.00
```

Table 1.20
Basic Number Alignment Examples

Option	Examples
Left-align within field (default)	`1` `2` ` ` ` ` ` ` ` ` `3` `.` `1` `4` ` ` ` `
Center-align within field	` ` ` ` `3` `5` ` ` ` ` ` ` `-` `1` `4` ` ` ` `
Right-align within field	` ` `3` `.` `1` `4` `1` `3` `.` `0` `+` `0` `8`

Right alignment with a given field width displays per below.

```python
for velocity in range(0, 50, 20):
  print(f"{velocity:>8.2f} {(velocity**2):>8.2f}")
```

```
    0.00      0.00
   20.00    400.00
   40.00   1600.00
```

The next shows using a space instead of a sign for positive numbers as detailed in Table 1.19.

```python
for angle in range(0, 360, 45):
    print(f'{math.sin(math.radians(angle)): }')
```

```
 0.0
 0.7071067811865475
 1.0
 0.7071067811865476
 1.2246467991473532e-16
-0.7071067811865475
...
```

1.10 IMPORTING MODULES

To use the functionality of existing Python modules, also called libraries or packages, first load them into the interpreter using an `import` or `from` statement. The `import` statement loads an entire module into the current namespace, which is a collection of names and their corresponding objects that can be accessed. The `from` statement is used to import selected objects within a module.

The syntax for an `import` statement is:

```python
import module_name
```

For example, the following `import` loads the `math` module into the current namespace. It can then be used as a prefix to its functions. For example, the value of the pi constant can be accessed using the dot notation.

```
import math
print(math.pi)
```

Alternatively use the `as` keyword to define a different alias name.

```
import module_name as alias_name
```

As an example the `random` library will be imported and given the name `rnd` with the `as` keyword. Its function to generate a random uniform number is called with the alias name using the dot notation.

```
import random as rnd
print(rnd.uniform(10, 20))
```

Frequently an alias name is defined as an abbreviation. For example, NumPy is usually imported as `np` and all its functions and objects referred to with the `np` name. It is highly recommended to stick with conventional names for overall compatibility and consistency.

The importing of frequently used general-purpose libraries overviewed in Chapter 2 and demonstrated in many examples are below. Shown are the conventional namespaces used for these libraries, but they could be named anything that is non-conflicting. For SciPy, specific submodules are generally imported, so there is no overall namespace used.

```
# general purpose library namespaces

# numpy for numerical computing with arrays
import numpy as np

# matplotlib for plotting and data visualization
import matplotlib.pyplot as plt

# pandas for data analysis
import pandas as pd

# scipy general purpose scientific computing statistics module
from scipy.stats import stats
```

The `from` statement can be used to import specific objects from a module. Its syntax is:

```
from module_name import object_name
```

This allows one to access the objects defined in a module without having to specify the full module name. For example, the following `from` statement imports the pi constant from the `math` module. Once it is imported, it can be accessed directly without having to write the `math` module prefix.

```python
from math import pi
print(pi)
```

The `import` and `from` statements are similar in that they both load modules into the current namespace. While the `import` statement imports all of the objects in a module, the `from` statement imports specific objects from a module.

1.11   READING INPUT FROM THE USER

The `input` function is used to read input from the user. There are two forms, with and without an argument:

- `input()` returns a string entered by the user.

- `input(prompt)` displays the `prompt` string, then returns the string entered by the user.

The following displays prompts and assigns the user inputs to variables:

```python
name = input("What is your name?")
print("Hello", name)
board_number = input("What is the circuit board serial number?")
What is your name? Ramón Ensayador
Hello Ramón Ensayador.
What is the circuit board serial number?
```

The value returned by `input` is always a string. To obtain and use numbers, the string must be converted to an integer or floating point value with the `int` or `float` functions respectively. This is demonstrated below. An exception will occur if numeric computations are attempted with a string. See Section 1.17 for handling exceptions with this example.

```python
number = float(input('What is the number to square? '))
print(number, 'squared = ', number * number)
What is the number to square? 5
5.0 squared =  25.0
```

1.12   READING AND WRITING FILES

Python provides built-in functions for creating, writing, and reading files. Access modes dictate the type of operations available with an opened file. The file may be read-only, write, append, read-write, or other modes as summarized in Table 1.21.

The contents of an entire file can be read with the `open` function. It opens a named file in read mode (the default mode if not specified) and returns a file object representing the opened file. The `read()` method is called to read the entire file content as a single string which is assigned to a variable.

Table 1.21

File Access Modes

Access Mode	Description
r	Opens a file for reading in text mode. End-of-file (EOF) is encountered when the internal pointer reaches the end of the file.
w	Opens a file for writing in text mode. Existing content is truncated, and if the file doesn't exist, it's created.
a	Opens a file for appending in text mode. The file is created if it doesn't exist. New data is written to the end of the file.
rb	Opens a file for reading in binary mode. Useful for handling non-text files like images or data.
wb	Opens a file for writing in binary mode. Existing content is truncated, and if the file doesn't exist, it's created.
ab	Opens a file for appending in binary mode. The file is created if it doesn't exist. New data is written to the end of the file.
x	Opens a file for exclusive creation. Fails if the file already exists. Useful for preventing accidental overwrites.
t	Opens a text file in text mode. This is the default mode for 'r' and 'w' on some systems.
b	Opens a binary file in binary mode. This is the default mode for 'rb' and 'wb' on some systems.
+	Opens a file for both reading and writing. The file must exist. Use 'r+' for reading and writing from the beginning of the file, or 'a+' for appending and reading/writing from the end.

```python
temperature_data = open("temperatures.csv").read()
```

A file can alternatively be iterated over line by line. The with statement is used for working with files to ensure proper handling of file resources, including automatically closing a file when there are exceptions during processing. The open function opens the file in read mode and returns a file object. The returned file object is assigned to the variable f with the as keyword. The for loop iterates over the lines present in the file object, where the current line is assigned to the variable line.

```python
# create lists of time and temperature measurements from a csv file
# initialize lists
times, temperatures = [], []
with open("temperatures.csv") as file:
    for line in file:
        # extract measurements from each line
        times.append(line.split(",")[0])
        temperatures.append(line.split(",")[1])
```

1.13   WRITING TO A FILE

Files can written to after specifying the write mode with basic file functions or additional libraries to handle file formatting. The writing can be done either all at once or line-by-line. The next example writes a multi-line string to a csv file in a single operation where the line endings in the string are purposely retained.

```python
temperature_data = """time, temperature, pressure
00:00, 235.3, 128.6
00:05, 237.1, 132.1
00:10, 237.0, 131.9
"""

# Open the file for writing
with open('temperature_data.csv', 'w') as file:
  # Write the entire string to the file
  file.write(temperature_pressure_data)
```

The next examples writes a csv file from data lists of timestamps, temperatures, and pressures which could be series of collected measurements or simulated data. It imports the csv package for working with CSV files, creates a list of consolidated values for each row, and writes them to a file all at once.

A list comprehension is used to iterate over the lists simultaneously using zip to create a new list `data_to_write`. Each element is a sub-list containing the corresponding time, temperature, and pressure (a row for the CSV). The CSV file is opened in write mode with the `'w'` access, a header row is written, then the `writerows` method of the CSV writer object is used to write everything in `data_to_write` at one time.

```python
import csv

# Data lists
times = ["00:00", "00:05", "00:10"]
temperatures = [235.3, 237.1 , 237.0]
pressures = [128.6, 132.1 , 131.9]

# Combine data into rows for writing
data_to_write = [[time, temperature, pressure] for time, ↵
    temperature, pressure in zip(times, temperatures, pressures)]

# Open the CSV file for writing in write mode (will overwrite ↵
    existing file)
with open('temperature_data.csv', 'w', newline='') as csvfile:
  writer = csv.writer(csvfile)

  # Write header row
  writer.writerow(["time", "temperature", "pressure"])

  # Write all data rows at once using writerows
  writer.writerows(data_to_write)
```

The previous examples use very short and simple data structures. More complex data and multiple formats can be handled with the exhaustive functions available in Pandas. See the Pandas Section 2.3 and subsequent examples.

1.14   ADVANCED FUNCTION TOPICS

1.14.1   RECURSIVE FUNCTIONS

A recursive function calls itself to solve certain types of problems. Sometimes data structures like trees and linked lists are traversed to evaluate many paths. A recursive function is used below to add up the weight of UAV system components described in a hierarchical dictionary of dictionaries representing three levels of decomposition. It traverses down the nesting hierarchy and calls itself again when a value is another nested dictionary. The lowest level components have weight values that are added up. In this case weights are at the second and third levels.

```python
def get_total_weight(system_dictionary):
    """Returns the total weight of all the components in a system."""

    total_weight = 0

    for key, value in system_dictionary.items():
        # If the value is a dictionary for another component, ⤶
            recursively call the get_total_weight() function
        if type(value) == dict:
            total_weight += get_total_weight(value)
        # If the key is "weight", add the value to the total weight
        elif key == "weight":
            total_weight += value

    return total_weight

UAV = {
  "Payload": {
    "Camera": {
      "weight": .7
    },
    "Thermal Sensor": {
      "weight": 1.2
    },
    "Lidar": {
      "weight": 1.5
    }
  },
  "Onboard Computer": {
      "weight": 0.6
  },
  ...
```

```
  "Power": {
    "Battery": {
      "weight": 3
    },
    "Harness": {
      "weight": 0.2
    }
  }
}
```

```
get_total_weight(UAV)
```

```
10.7
```

1.14.2 FUNCTIONS AS INPUTS TO FUNCTIONS

A powerful feature is the ability to send entire functions as input to other functions. This can make for clean, elegant, reusable code with logical partitioning of the problem. For example, it is desired to have a general method for numerical integration instead of writing separate integration algorithms each time for various derivative functions. A single subroutine can be written that will integrate any time derivative function provided to it.

Euler's simple method of integration will be used with time as the independent variable. It computes the following at each time increment per Equation 1.1:

$$y(t_{n+1}) = y(t_n) + \frac{dy}{dt}(t_n)\Delta t \tag{1.1}$$

where

y is a function over time
$\frac{dy}{dt}$ is the time derivative of y
n is the time index
Δt is the time step size to increment t_n to t_{n+1}.

The routine in Listing 1.1 will accept any arbitrary derivative function with its own set of variables. This generality is afforded by the unpacking of arguments and keyword arguments with the $*$ and $**$ syntax. The signature of the function never has to change for new derivatives to accommodate different parameters.

It is demonstrated with two separate functions. It passes the differential equation for Newton's law of cooling to the `euler` function to integrate and secondly models exponential growth. Parameters specific to each derivative function are sent as inputs.

Listing 1.1 Function Inputs to Functions

```
def euler(f, y0, t0, tf, dt, *args, **kwargs):
    """

    Performs numerical integration using the Euler method.

    Args:
```

```
        f (function): A function that returns the derivative of `y` ↙
            with respect to time.
        y0 (float): The initial value of `y`.
        t0 (float): The start time of the integration.
        tf (float): The end time of the integration.
        dt (float): The step size for the integration.
        *args: Additional positional arguments to pass to the ↙
            derivative function `f`.
        **kwargs: Additional keyword arguments to pass to the ↙
            derivative function `f`.

    Returns:
        list of tuple: A list of tuples where each tuple contains a ↙
            time `t` and the corresponding value of `y(t)`.
    """

    time, y = t0, y0
    print("time y")
    while time <= tf:
        print(f"{time:.1f} {y:.1f}")
        y += f(y, *args, **kwargs) * dt
        time += dt

def newton_cooling(temp, external_temp, coefficient):
    """
    Applies Newton's Law of Cooling to calculate the rate of ↙
        temperature change.

    Args:
        temp (float): Current temperature of the object.
        external_temp (float): Ambient or external temperature.
        coefficient (float): Heat transfer coefficient for the rate ↙
            of heat loss.

    Returns:
        float: The rate of temperature change.
    """
    return -coefficient * (temp - external_temp)

def exponential_growth(x, rate):
    """
    Models exponential growth by multiplying the current value by a ↙
        constant growth rate.

    Args:
        x (float): The current value.
        rate (float, optional): The growth rate.

    Returns:
```

```
        float: The value after applying exponential growth.
    """
    return x * rate

print("Newton Cooling from 100°")
euler(newton_cooling, y0=100, t0=0, tf=50, dt=1, coefficient=0.07, ↙
    external_temp=20)

print("Exponential Growth from 10")
euler(exponential_growth, 10, 0, 5, 1, rate=1.3)
```

```
Newton Cooling from 100°
time  y
0.0 100.0
1.0 94.4
2.0 89.2
3.0 84.3
4.0 79.8
...
Exponential Growth from 10
time  y
0.0 10.0
1.0 23.0
2.0 52.9
3.0 121.7
```

1.14.3 DOCSTRINGS

When docstrings are present in the beginning of a function, module, class, or method, they are associated with the object as a __doc__ attribute. Docstrings can be useful when one desires help on a function from an imported library that is not immediately visible. Suppose one wants to check the usage of the math sine function. One can get help information via the `help`(math.sin) command in the interpreter or programmatically as below.

```
print(math.sin.__doc__)
```

```
Return the sine of x (measured in radians).
```

A minimum docstring can be embedded with triple quotes and automatically extracted.

```
def projectile(v0, angle):
    """ Returns the projectile flight time, maximum height, and ↙
        distance given initial velocity in meters per second and ↙
        launch angle in degrees. """

    g = 9.8# gravity (meters per second squared)
    angle_radians = 0.01745 * angle # convert degrees to radians
```

```
flight_time = 2 * v0* math.sin(angle_radians) / g
max_height = (1 / (2 * g)) * (v0 * sin(0.01745 * angle)) ** 2
distance = 2 * v0** 2 / g * sin(angle_radians) * cos(angle_radians)
return(flight_time, max_height, distance)
```

After the function is already imported or read into the interpreter, the docstring can be accessed with the `__doc__` method:

```
projectile.__doc__
```

```
Returns the projectile flight time, maximum height, and distance
given initial velocity in meters per second and launch angle in
degrees.
```

A more complete docstring will specify input parameters and return parameters as below. With this the `__doc__` method will return everything in the triple quotes.

```
def projectile(v0, angle):
    """
    Calculates the flight time, maximum height, and distance of a ↵
        projectile.

    Args:
        v0 (float): Initial velocity of the projectile in meters per ↵
            second.
        angle (float): Launch angle of the projectile in degrees.

    Returns:
        tuple: A tuple containing:
            - flight_time (float): Projectile flight time in seconds.
            - max_height (float): Maximum height reached by the ↵
                projectile in meters.
            - distance (float): Horizontal distance traveled by the ↵
                projectile in meters.
    """
    ...
```

Many tools and environments parse the docstrings to generate help and auto-mated documentation in multiple formats. For example in smart editors, when hovering over or typing in a function (or class) name, the documentation is displayed as a tooltip popup. Figure 1.2 demonstrates a tooltip that displays with the cursor focused on the `projectile()` function statement. See further material on docstrings and automated documentation in Chapter 5.

1.14.4 LAMBDAS

A lambda is a small, anonymous function defined by a single expression. It serves as a function shortcut which will evaluate a single expression and return the result. Unlike other functions, lambda functions don't have names or docstrings.

```
def projectile(v0, angle)
```

Open in tab   View source

Calculates the flight time, maximum height, and distance of a projectile.
Args:
    v0 (float): Initial velocity of the projectile in meters per second.
    angle (float): Launch angle of the projectile in degrees.
Returns:
    tuple: A tuple containing:
    - flight_time (float): Projectile flight time in seconds.
    - max_height (float): Maximum height reached by the projectile in meters.
    - distance (float): Horizontal distance traveled by the projectile in meters.

Figure 1.2   Function Docstring Tooltip Example

A lambda can be assigned to a variable or passed as an argument to another function. Lambda functions are often used in conjunction with higher-order functions, which take one or more functions as arguments or return one or more functions. Lambda functions can help to simplify code and make it more concise with less overhead than a full function.

The syntax requires the `lambda` keyword with input parameters and an expression. The parameters are a sequence of variables that will be used in the expression.
Lambda Syntax

```
lambda [parameters]: expression
```

The following lambda function takes two arguments for mass and velocity and returns the kinetic energy. The function will be bound to the values of the arguments when it is called, the expression will be evaluated, and the result will be returned as follows:

```
kinetic_energy = lambda mass, velocity: mass * velocity / 2
kinetic_energy(50,30 ) # 50 kilograms, velocity 30 m/s
750.0
```

Lambda functions can also be used as arguments to other functions and composed with iterables. The following example calculates the portion of projectile launcher targeting errors that are less than 2 meters. The lambda function `lambda x : x<2` checks if a value is less than 2, and returns `True` if it is or `False` otherwise. The `map` function applies the lambda function to each element in the list, creating another iterator containing `True` or `False` elements. The `sum` function adds up the values in the iterator, counting the number of `True` values corresponding to measurements less than 2.

```
# determine fraction of target errors less than 1 meter
target_errors = [1.2, 0.0, 3.2, .8, 0.7, 0.4, 1.6, 2.7, 1.3, 3.1, .5↙
    , 0.2, 1.0, 2.8, 1.3, .5, 1.8]
sum(map(lambda x : x<2, target_errors)) / len(target_errors)
0.7647058823529411
```

1.15 CLASSES AND OBJECT-ORIENTED PROGRAMMING

Object-Oriented (OO) programming organizes software around the concept of objects, which represent real-world entities, and their interactions. Classes are used to define objects. A class is like a template or blueprint from which objects are created. An object is called an *instance* of a class. Python supports all major OO concepts. The paradigm is highly prevalent in the popular libraries used.

A class defines the name, attributes, and methods of an object. Attributes are variables that hold information about an object instance. They describe the state of an object. Attributes can be accessed and modified by code outside of the class. Methods are functions defined within a class describing object behavior. They can modify the object attributes and are called like functions. When objects are instantiated from a class (also called a *constructor*), they are given names and their attributes and methods can be referred to with the dot notation.

A *class diagram* in UML is useful to understand and design object-oriented software. It visualizes the structure of a system's classes, their attributes, operations, and relationships among objects. A class is represented by a rectangle divided into three compartments, displaying the class name, attributes, and operations as in Figure 1.3 for a simple car class. It is straightforward to elaborate the visualization into a working program as illustrated next.

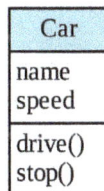

Figure 1.3   Class Diagram

This next example of a car class corresponds to the class diagram in Figure 1.3. A car object has attributes for name and speed, and methods to drive and stop. A default __init__ method initializes the attributes that describe each car using the dot notation. Conventionally *self* is used as the internal name for an object though it could be anything.

Attributes are defined within the class body, either directly assigning a value or using methods to set or retrieve them. Attributes that are unique to each object instance are defined and modified using the self.attribute_name dot notation.

Class attributes that are shared by all objects of the class are defined without using
`self`.

Methods are defined within the class body using the `def` keyword. They take
arguments typically including the special `self` argument that refers to the current
object instance. Methods can access and modify the object's attributes.

```python
class Car:
  def __init__(self, name):
    self.name = name
    self.speed = 0

  def drive(self, speed):
    self.speed = speed
    print(f"The {self.name} is driving at a speed of {self.speed} ↲
        miles per hour")

  def stop(self):
    print(f"The {self.name} has stopped.")
```

To create a car object, its name and speed are passed to the class as `Delorean` ↲
`= Car("Delorean", 90)`. Once it is created, its attributes can be accessed and methods can be called. E.g., For example, call the `Delorean.drive()` method to make the
car drive.

```python
# Create a Car object
Delorean = Car("Delorean")

# Call the drive() method with a speed
Delorean.drive(90)

# Call the stop() method
Delorean.stop()
```
```
The Delorean is driving at a speed of 90 miles per hour
The Delorean has stopped.
```

Previously the projectile launcher examples were written using a functional, or
imperative programming approach. Next in Listing 1.2, it is recast to be object-
oriented using a class to represent a projectile. It defines the projectile class attributes
and methods with additional code beyond a strictly functional approach. The methods are nearly identical to previous trajectory functions, but internally they require a
`self` parameter to associate them to a specific object.

Listing 1.2   Projectile Class

```python
import math

class Projectile:
    """
```

```
A class representing a moving projectile that calculates its ↙
    trajectory.

Attributes:
    angle (float): The initial angle of the projectile in degrees.
    velocity (float): The initial velocity of the projectile in ↙
        meters per second.
    g (float): The acceleration due to gravity in meters per ↙
        second squared.

Methods:
    calculate_trajectory(time_step: float, end_time: float) -> ↙
        list[tuple[float, float]]:
        Calculates the trajectory of the projectile over a ↙
            specified time period with a given time step.
"""


def __init__(self, angle: float, velocity: float, g: float = 9.81):
    """
    Initializes a projectile object with the specified angle, ↙
        velocity, and acceleration due to gravity.

    Args:
        angle (float): The initial angle of the projectile in ↙
            degrees.
        velocity (float): The initial velocity of the projectile ↙
            in meters per second.
        g (float, optional): The acceleration due to gravity in ↙
            meters per second squared. Defaults to 9.81.
    """
    self.angle = angle
    self.velocity = velocity
    self.g = g

def launch(self, time_step: float, end_time: float) -> ↙
    list[tuple[float, float]]:
    """
    Launch and calculate the trajectory of the projectile over ↙
        time.

    Args:
        time_step (float): The time step in seconds.
        end_time (float): The end time in seconds.

    Returns:
        A list of tuples representing the (x, y) coordinates of ↙
            the projectile at each time step.
    """
    # Convert the angle to radians
```

```python
        angle_rad = math.radians(self.angle)

        time_step = 1

        # Calculate the initial horizontal and vertical velocities
        v0x = self.velocity * math.cos(angle_rad)
        v0y = self.velocity * math.sin(angle_rad)

        # Initialize the x and y coordinates and velocities
        x = 0
        y = 0
        vx = v0x
        vy = v0y

        time = 0

        # Initialize the trajectory list with the initial time and ↲
            coordinates
        trajectory = [(time, x, y)]

        print("The projectile is launched")

        # Iterate over time steps and calculate the new coordinates ↲
            and velocities
        while y >= 0:
            time += time_step
            # Calculate the new x and y coordinates
            x = x + vx * time_step
            y = y + vy * time_step - 0.5* self.g * time_step**2

            # Calculate the new velocities
            vx = v0x
            vy = v0y - self.g * time * time_step

            # Add the new coordinates to the trajectory list
            trajectory.append((time, x, y))

        return trajectory
```

It is used by first instantiating a projectile object giving a name, velocity, and angle. The launch method is called the output trajectory and is returned as a list of tuples containing the time, x, and y coordinates.

```python
ball = Projectile(angle = 45, velocity = 100)
ball.launch()

The projectile is launched
[(0, 0, 0),
 (1, 70.71067811865476, 65.80567811865474),
 (2, 141.4213562373095, 121.80135623730948),
```

```
(3, 212.13203435596427, 167.9870343559642),
(4, 282.842712474619, 204.36271247461895),
(5, 353.5533905932738, 230.9283905932737),
(6, 424.26406871192853, 247.68406871192843),
(7, 494.9747468305833, 254.62974683058317),
(8, 565.685424949238, 251.76542494923794),
(9, 636.3961030678928, 239.09110306789267),
...
```

1.15.1 INHERITANCE

Inheritance is the creation of new classes (subclasses) that inherit properties and functionalities from existing classes (parent classes). The subclass inherits all attributes and methods of the parent class by default. These can be overridden, or new attributes and specialized behavior can be added. Inheritance promotes code reusability and establishes a hierarchical relationship between classes.

The following example expands the Car class, making it a parent class for electric and gas cars. It introduces a class-level attribute for distance and adds specific parameters for gas and electric cars. The distance traveled is calculated using the time and speed inputs. Inheritance is implemented using the class ↙ SubclassName(SuperclassName) syntax.

The class diagram in Figure 1.4 shows the inheritance relationships for this example. In UML notation, a solid line with a closed arrowhead points to the parent class, indicating inheritance. Corresponding to the code, the ElectricCar and GasCar classes inherit properties from the parent class while also adding additional attributes and methods to distinguish the object types.

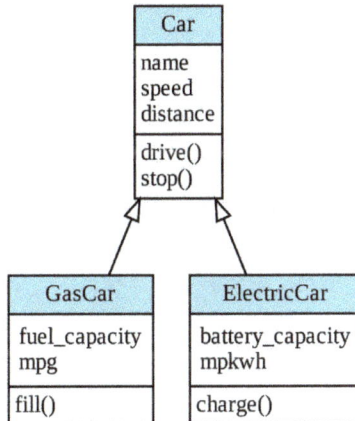

Figure 1.4 Class Diagram with Inheritance

Listing 1.3   Class Inheritance

```python
class Car:
  def __init__(self, name, speed):
    self.name = name
    self.speed = speed
    self.distance = 0

  def drive(self, time):
    self.distance += self.speed * time
    print(f"The {self.name} is driving at a speed of {self.speed} ↵
        for {time} hours")

  def stop(self):
    print(f"The {self.name} has stopped after {self.distance} miles.")

class ElectricCar(Car): # ElectricCar inherits from Car
  def __init__(self, name, speed, battery_capacity, mpkwh):
    super().__init__(name, speed)
    self.battery_capacity = battery_capacity # battery capacity in ↵
        Kilowatt hours)
    self.mpkw = mpkwh # Miles Per Charge

class GasCar(Car): # GasCar inherits from Car
  def __init__(self, name, speed, fuel_capacity, mpg):
    super().__init__(name, speed)
    #self.distance = 0
    self.fuel_capacity = fuel_capacity
    self.mpg = mpg # Miles Per Gallon

# Create a GasCar object
Jeep = GasCar("Jeep", speed=80, fuel_capacity=22, mpg=16)

# Create a ElectricCar object
Tesla = ElectricCar("Tesla", speed=90, battery_capacity=80, mpkwh=4)

# Call the drive() method
Jeep.drive(time=10)
Tesla.drive(time=8)

# Call the stop() method
Jeep.stop()
Tesla.stop()

The Jeep is driving at a speed of 80 for 10 hours
The Tesla is driving at a speed of 90 for 8 hours
The Jeep has stopped after 800 miles.
The Tesla has stopped after 720 miles.
```

1.15.2   AGGREGATION

Aggregation between classes describes a "part-of" relationship, where a class uses the functionality of other classes. The class diagram in Figure 1.5 illustrates aggregation between classes for a UAV. The UML notation uses a hollow diamond at the end of an association, indicating this whole-part relationship. The UAV is composed of an engine and a battery, both represented as separate classes. The battery and engine are considered "part-of" the UAV.

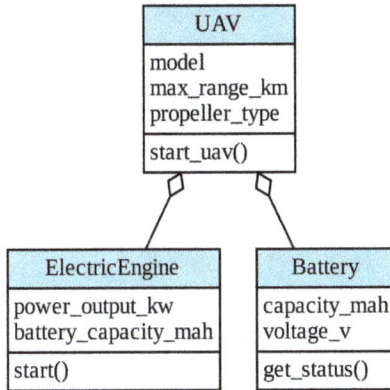

Figure 1.5   Class Diagram with Aggregation

The corresponding example of aggregation for the UAV class is in Listing 1.4. The ElectricEngine and Battery classes represent independent objects with their own properties and methods. The UAV class represents a more complex object that is composed of an ElectricEngine and Battery, demonstrating aggregation. The ElectricEngine and Battery are passed to the UAV during initialization and become part of its structure.

Listing 1.4   Class Aggregation

```python
class ElectricEngine:
    def __init__(self, power_output_kw, battery_capacity_mah):
        self.power_output_kw = power_output_kw
        self.battery_capacity_mah = battery_capacity_mah

    def start(self):
        return f"Electric engine started with {self.power_output_kw} ⤶
            kW power and {self.battery_capacity_mah} mAh battery ⤶
            capacity."


class Battery:
    def __init__(self, capacity_mah, voltage_v):
```

```python
        self.capacity_mah = capacity_mah
        self.voltage_v = voltage_v

    def get_status(self):
        return f"Battery capacity: {self.capacity_mah} mAh, Voltage: ↙
            {self.voltage_v}V"


class UAV:
    def __init__(self, model, max_range_km, engine, battery, ↙
        propeller_type):
        self.model = model
        self.max_range_km = max_range_km
        self.engine = engine # Aggregation: UAV "has an" Electric ↙
            Engine
        self.battery = battery # Aggregation: UAV "has a" Battery
        self.propeller_type = propeller_type

    def start_uav(self):
        engine_status = self.engine.start()
        battery_status = self.battery.get_status()
        return (f"UAV {self.model} (Max Range: {self.max_range_km} ↙
            km, Propeller: {self.propeller_type}):\n"
                f"{engine_status}\n{battery_status}")


# Create instances of ElectricEngine and Battery
engine = ElectricEngine(2.5, 5000) # 2.5 kW power output, 5000 mAh ↙
    battery capacity
battery = Battery(5000, 14.8) # 5000 mAh capacity, 14.8V

# Create a UAV instance with the ElectricEngine and Battery instances
uav = UAV("Phantom 4", 20, engine, battery, "Fixed-pitch")

print(uav.start_uav())
```

```
UAV Phantom 4 (Max Range: 20 km, Propeller: Fixed-pitch):
Electric engine started with 2.5 kW power and 5000 mAh battery capacity.
Battery capacity: 5000 mAh, Voltage: 14.8V
```

1.15.3   ASSOCIATION

The example in Listing 1.5 combines the previous Car class with a new Race class. It demonstrates how any number of car instances can be added to a race, illustrating a class *association* relationship. Class associations are used to model relationships where objects of one class are linked to objects of another. The Race class has a method to add Car objects along with their attributes, modeling participants in an actual race environment.

As shown in the class diagram in Figure 1.6, an association between classes is represented by a solid line connecting them, indicating the presence of a relationship. Multiplicity values at the ends of the connection illustrate how many instances of one class can be associated with instances of another. In this case, the multiplicity indicates that a single `Race` object can be associated with one or more instances of the `Car` class, represented by the notation `1..*`. This means that a race must have at least one car but can include multiple cars.

Associations can be classified into different types, such as one-to-one, one-to-many, and many-to-many. The multiplicity notation (`1..*` in this example) helps specify these types explicitly. These associations provide clarity in class design so the software accurately reflects the intended relationships, enhancing modularity and maintainability.

Inheritance, association, aggregation, and other UML properties can be combined within a class diagram to create more sophisticated and realistic representations of complex systems. When used together, they allow for modeling the hierarchical structure, interactions, and dependencies between different components. Such diagrams provide a comprehensive and interconnected view of the system, enabling better planning, analysis, and communication.

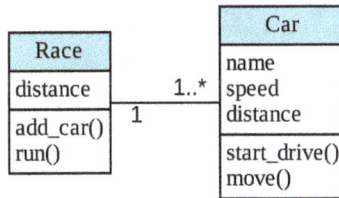

Figure 1.6   Class Diagram with Association

Listing 1.5   Class Association

```python
import random
import time

# Time interval
dt = .1

class Car:
    def __init__(self, name):
        self.name = name
        self.speed = 0
        self.distance = 0 # To track how far the car has gone

    def start_drive(self, speed):
        self.speed = speed
```

```python
        print(f"The {self.name} is driving at a speed of {self.speed} ↵
            miles per hour.")

    def move(self):
        # Move the car based on its speed
        self.distance += self.speed * dt

class CarRace:
    def __init__(self, race_distance):
        self.cars = []
        self.race_distance = race_distance

    def add_car(self, car):
        self.cars.append(car)
        print(f"{car.name} has been added to the race.")

    def run(self):
        print(f"The race has started. The distance is ↵
            {self.race_distance} miles.\n")

        # Randomly set each car's speed at the start of the race
        for car in self.cars:
            car.start_drive(random.randint(60, 90))

        # Simulate the race until at least one car reaches the target ↵
            distance
        finished = False
        time = 0
        #dt = 1
        while not finished:
            time += dt
            for car in self.cars:
                car.move()

                # Check if any car has reached or exceeded the race ↵
                    distance
                if car.distance >= self.race_distance:
                    finished = True

        # Find all cars that reached or exceeded the race distance
        winners = [car for car in self.cars if car.distance >= ↵
            self.race_distance]
        if len(winners) == 1:
            print(f"\nThe race is over at time {time:.1f} hours. The ↵
                winner is {winners[0].name}.")
        else:
            print(f"\nThe race is over at time {time:.1f} hours. It's ↵
                a tie between:")
            for winner in winners:
```

```
                    print(f"{winner.name}")

# Create cars
delorean = Car("Delorean")
fiat = Car("Fiat")

# Create a race
race = CarRace(race_distance=500)
race.add_car(delorean)
race.add_car(fiat)

# Run the race
race.run()
```

```
Delorean has been added to the race.
Fiat has been added to the race.
The race has started. The distance is 500 miles.

The Delorean is driving at a speed of 71 miles per hour.
The Fiat is driving at a speed of 80 miles per hour.

The race is over at time 6.3 hours. The winner is Fiat.
```

1.16   GENERATORS

Generators are powerful functions for creating iterators. They produce values or objects one at a time without creating entire lists in memory. They are more efficient this way to generate sequences of values on the fly compared to list comprehensions or `for` loops. They only generate the next item when it is needed (also called *lazy*), rather than all at once.

Generators are appropriate for handling very large datasets to save memory, computationally expensive operations, or when the sequence length is unknown or infinite. Data-intensive applications can exceed memory limits. The number of required values to generate may be unknown; for example in event-driven simulations events may spawn new random events, or similarly in agent-based simulations with emergent behaviors

There are two ways to create objects that generate values typically consumed in a `for` loop: 1) a generator function that uses the `yield` statement to return values or 2) a generator expression inside parentheses.

The `yield` statement defines a generator function. It returns a value or object to the caller like a `return` statement, then the function pauses execution until it is called again and resumes at the same line. This works differently than traditional functions that always start from their beginning when called. This can be non-intuitive when using `yield` and initially difficult to follow.

The following `launch_countdown` function creates a generator of fixed length. It will return a value of time from line 4 whenever it is called from the main loop in line

8 (while time is valid). When it is invoked next it will proceed to line 5 to decrement the time and continue the `while` loop.

```python
1 # Generator function to yield countdown time
2 def launch_countdown(time):
3   while time > 0:
4     yield time
5     time -= 1
6
7 print('T-Minus')
8 for time in launch_countdown(10):
9   print(time)
10 print('and we have liftoff!!')
T-Minus
10
9
8
...
2
1
and we have liftoff!
```

A generator expression is similar to a list comprehension, but used only for iteration without constructing a list in memory.

Generator Expression Syntax

```python
(expression for item in sequence if condition)
```

The generator expression is equivalent to a `for` loop with a `yield`.

```python
for item in s:
  if condition:
    yield expression
```

The parentheses on a generator expression can be dropped if used as a single function argument, e.g. `sum(x*x for x in s)`.

An equivalent generator expression for the launch countdown is straightforward.

```python
launch_countdown = (time for time in range(10, 0, -1))

for time in launch_countdown:
    print(time)
```

Generator expressions can allow for much less code to be written compared to traditional `for` loops. Furthermore, sequences of generator expressions that pipeline to each other can replace nested `for` loops. They also eliminate creating lists that consume memory that could be constraining otherwise.

This example computes the average in-production time of circuit cards per day after opening a very large log file of circuit card production data. Each line contains an electronic serial number, automated timestamps for each production step, and the last column is the total process time in minutes for successfully tested cards. The last column is blank for test failures and must be filtered out of the calculations.

```
E236-FZAB-4597, 10-12-2023 9:47:56, ..., 214
E714-SMIL-3612, 10-12-2023 9:52:28, ..., 196
E149-BURB-4761, 10-12-2023 9:54:05, ...,
...
```

The following two generator expressions perform the same as nested `for` loops to read each line with an underlying conditional check for valid values. The first `seconds` generator yields seconds by splitting each line using a comma delimiter to access the last column data. The `production_time` generator creates integer values from the `seconds` generator only for successful cards in a pipelining approach.

```python
from statistics import mean

with open("circuit_cards.csv") as production_data:
    seconds = (line.strip().split(",")[-1] for line in production_data)
    production_time = (int(x) for x in seconds if x != '')
    print("Mean Production Time", mean(production_time))
```

Generators are versatile, and their outputs can be accessed in different ways. They can be used to generate sequences of a specified length, indeterminate or infinite. They generate numeric values or complex objects which can be accessed in iterator `for` loops or with the `next` method.

An example of a long time horizon simulation with non-deterministic event times is a nuclear decay process. The following infinite generator function yields the time interval until the next decay event as an exponentially distributed random number (corresponding to a Poisson distribution for the number of events per time interval). It also shows how the `next` method can be used to access the next generated succcessive value, infinitely if desired.

```python
from numpy import random

def generate_nuclear_decay_time(mean):
  while True:
    yield random.exponential(mean)

# Create generator object for mean time between decay events of .03
generator = generate_nuclear_decay_time(mean=.03)

time = 0
while time < 100: # 100 years
    # Get the next value from the generator
    next_decay_time = next(generator)
    print(next_decay_time)
    time += next_decay_time
```

```
0.03631109190180292
0.023005702411883964
0.005083286542328262
...
```

More can be accomplished with generators. They can be used to read and process very large datasets in chunks rather than loading the entire dataset into memory at once. They can be composed with other generators or iterable objects using functions such as `zip()` and `map()` for powerful data manipulation and analysis.

1.17   EXCEPTIONS

An *exception* is a signal from the Python interpreter during runtime that an error or other unusual condition has occurred. There are numerous built-in exceptions, and new ones can be defined.

1.17.1   RAISING EXCEPTIONS

Python will raise an exception when a program attempts to do something erroneous. Below an error occurs when the input data list is blank leading to a zero in the denominator.

```
>>> mean = sum(temperature_data) / len(temperature_data)
-----------------------------------------------------------------
Traceback (most recent call last):

  File "/var/folders/nb/lqpqgz253d7bh24zdjqk_vvr0000gq/T/
  ipykernel_43145/1806623527.py", line 1, in <module>
    mean = sum(input_data) / len(input_data)

ZeroDivisionError: division by zero
```

This traceback indicates that the `ZeroDivisionError` exception is being raised. This is a built-in exception.

1.17.2   CATCHING EXCEPTIONS

Exception handling blocks can be set up to catch errors using the keywords `try` and `except`. When an error occurs within the `try` block, the interpreter looks for a matching `except` block to handle it. If there is one, execution jumps there. The following handler will catch a `ZeroDivisionError` exception in the calculation and inform the user.

```python
temperature_data = []
try:
    mean = sum(temperature_data) / len(temperature_data)
except ZeroDivisionError:
    print("You can't divide by zero. The input data is blank.")
```

```
You can't divide by zero.  The input data is blank.
```

The next example will catch non-numeric input for calculating a square number and inform the user to re-input until it is valid. Without the handler a `ValueError` would result if the input was not numeric.

```python
# Trap invalid number input to avoid program crash and inform user.
# Continue loop until it is valid then break out of loop.
while True:
  number = input('What is the number to square? ')
  try:
    number = float(number)
    break
  except ValueError:
    print ("That is not a floating point number. Try again")

number_squared = number * number
print(number, 'squared = ', number_squared)
```

If you don't specify an exception type on the except line, it will catch all exceptions. See https://docs.python.org/3/tutorial/errors.html for further information on errors and exceptions.

1.18   SUMMARY

Python style and conventions make programming easier. Code should be written in a consistent and readable fashion, using indentation appropriate to the logic to make it understandable.

Variables are used to store data and can be assigned any valid data type. Variable names should be descriptive and meaningful. Data types include strings, numbers, lists, dictionaries, tuples, sets, and Booleans. A variety of operators can be used on data for arithmetic operations, assignment operations, comparison operations, logical operations, identity operations, and membership operations.

If statements and conditions are used to govern program logic. If statements control the flow of a program by executing different blocks of code based on the evaluation of a condition. Short-hand if, else, elif, and short-hand if ... else statements can be used to make code more concise. Nested conditionals can be used to create more complex branching logic.

Data structures and their methods provide many flexible options for representing and processing engineering data. Lists are mutable data structures used to store collections of ordered items of any (mixed) type.

Dictionaries are a mutable data structure that can be used to store a collection of key-value pairs. They are a powerful data structure for storing and accessing data in a logical, organized manner. They can serve as containers for all parameters and variables in an analysis.

Tuples are an immutable data structure that can be used to store a collection of ordered items similar to lists. They are preferable to lists for efficiency when the data sequences do not change after initial creation. Sets are a mutable data structure that can be used to store collections of unique items.

Iterations of repeated operations can be performed with `for`, `while`, and `enumerate` loops. `for` loops are used to iterate over a collection of data items in data structures, executing a block of code for each item. `while` loops can be used to execute a block of code repeatedly until a logical condition is met. `enumerate` is similar to a `for` loop but returns a tuple containing a numeric index with each item.

Number and string formatting can be accomplished several ways. F-strings are the most natural and powerful method. The string `format` method and more primitive string formatting can also be used.

Basic utilities for data input include the `input()` function to read input from the user and file reading functions. Print statements are used to output data to the console, and files can be written to in various access modes.

Importing of modules to use existing code functionality in a program is common and frequently necessary.

Functions can be used to encapsulate code and make it reusable. They can take arguments with or without keywords and return values. Expressions and functions can be nested. Recursive functions, functions as inputs to functions, and function docstrings can be used to make functions more powerful, flexible, and usable.

Python can be written with imperative programming functions or object-oriented classes. A class is like a cookie cutter template representing collections of real-world objects that are instantiated. Classes are used to create templates for objects with associated attributes and methods. Objects are instances of the classes created at run time.

Inheritance, association, and aggregation are important relationships in object-oriented modeling. These relationships can be combined within a class diagram to create more sophisticated and realistic representations of complex systems. When used together, they allow for modeling the hierarchical structure, interactions, and dependencies between different components.

Generators provide values on the fly and are memory efficient for large iterables. They are useful for managing large data sets, computing expensive operations efficiently, and generating complex sequences of values.

Exceptions can be used to handle errors in code. They allow one to gracefully handle errors and continue program execution..

1.19 GLOSSARY

argument: A value passed to a function or method when it is called. There are two kinds:

> **keyword argument:** An argument that is passed to a function by name, preceded by a keyword identifier in a function call (e.g., `name=value`), or passed in a dictionary preceded by `**`.

> **positional argument:** An argument that is passed to a function by position without a keyword. Positional arguments can appear at the

beginning of an argument list and/or be passed as elements of an iterable preceded by *.

assignment: A statement that assigns a value to a variable using the equals sign =.

Boolean: A data type with two possible values of `True` or `False`.

class: A template for creating objects. A class defines a set of attributes and methods that the created objects will have. Classes are defined using the `class` keyword.

comment: Line(s) in a program not executed and used to provide explanations or notes. Single line comments begin with a hash mark # and multiline comments are surrounded by triple quotes.

data type: A category of data items, such as integers, floats, strings, and Booleans, that defines the kind of operations that can be performed on the data.

dictionary: An unordered collection of key-value pairs that is mutable, and can be changed after creation. Dictionaries are defined by curly braces {} and use keys to map to their corresponding values. Each key in a dictionary must be unique.

exception: An error that is detected while a program is running, often causing the program to halt unless the error is handled.

expression: A combination of variables, operators, and values that produces a single result.

function: A block of reusable code that performs a specific task. Functions are defined using the `def` keyword and can take arguments to operate on and return data.

immutable: A data object whose state cannot be modified after it is created including tuple data structures.

indentation: Leading whitespace (spaces or tabs) at the beginning of a line of code used to define the structure and hierarchy of code blocks.

indexing: The process of accessing an element of a sequence, such as a list, tuple, or string, using its position or index. Indexing starts at zero.

iteration: The process of looping through elements in a sequence or repeatedly executing a block of code.

iterable: An object capable of returning its members one at a time, allowing it to be looped over in a `for` loop.

keyword: A reserved word that is used by the interpreter to parse a program (as opposed *keyword* referring to key names in *keyword arguments*). Keywords are in Appendix Table A.1.

lambda: An anonymous, inline function defined using the `lambda` keyword.

list: An ordered collection of items that is mutable, which can be changed after it is created. Lists are defined by square brackets [] and can store elements of different data types.

literal: A notation for representing a fixed value and data type.

method: A function that is defined inside a class and is associated with an object. Methods are used to perform operations on objects of that class.

mutable: A data object whose state or contents can be modified after it is created including lists and dictionaries.

object: An instance of a class. Objects are created using a class definition and have attributes (data) and methods (functions) defined by the class.

operator: A symbol that represents a computation or operation, such as addition +, subtraction -, or comparison ==.

set: An unordered collection of unique items that is mutable. Sets are defined by curly braces {} without key-value pairs.

slicing: A mechanism for extracting a portion of a sequence, such as a list, tuple, or string, using a specific range of indices. Slicing is performed using the colon : operator within square brackets [].

statement: An instruction that executes a command or action including assignment statements and control statements.

string: A sequence of characters enclosed in quotes (single, double, or triple) used to represent text.

tuple: An ordered collection of items that is immutable, whose elements cannot be changed after it is created. Tuples are defined by parentheses () and can store elements of different data types.

variable: A name that refers to a value stored in memory used to hold data.

1.20   EXERCISES

1. Use the general Euler's integration routine for a given scenario or improve it. For example, use it to calculate the height over time of a projectile by passing its vertical velocity as the derivative. Alternatively, replace it with a more accurate Runge-Kutta algorithm, and test it with existing examples and new derivative functions.

2. Write a recursive function that accepts a nested dictionary and prints out a summary of each dictionary it encounters. It should start at the top level and indicate the level of nesting for each dictionary in the printout.

3. Extend the recursive function from the previous exercise to parse nested dictionaries representing a system tree structure. Each dictionary corresponds to a node, and nested dictionaries represent branches. For each dictionary, report its level and the number of branch connections. Output the total number of levels, and order the dictionaries by their number of branch connections.

4. Develop a routine that accepts natural language input for spelled-out numbers and performs the appropriate conversions (e.g., "two" to 2), as an AI chatbot or search engine would handle inputs. Use this routine as a generalized front-end for other computations.

5. Create another engineering analysis example involving passing functions as arguments to other functions.

6. Create a dictionary that stores the physical properties of materials, and write functions to calculate derived properties for a given material. For example, it could take a material's name and calculate the mass of a given volume after retrieving its density. Implement error handling in case the material is not found in the dictionary.

7. Write a program that accepts chemical reaction formulas as strings (e.g., $H2$⦰ + $O2$ -> $H2O$) and checks if the equation is balanced by comparing the number of atoms on each side. Use string parsing and dictionaries to process the elements and their counts.

8. Write a class, ControlSystem, for a simple finite state machine that simulates a control system with three states: Idle, Active, and Error. Logical transitions between states depend on input conditions. Handle input signals for Start, Error, and Reset. Optionally, add more states or extend the system for a specific control application.

9. Write a program that displays consecutive data values vertically as text on separate lines, where the number of characters per line is scaled according to the value. For example, a value that is half of the maximum would display 40 characters on an 80-character line. Commonly, asterisks are used. Input the data values as separate lists or tuples.

10. Extend the previous text display program to plot multiple variables on each row. Assign a different text symbol (which can be colored) for each parameter, and plot each value as a marker on the scaled line. Generate a set of multi-parameter data over time and plot it.

11. Write a function that reads sensor data from a CSV file, processes it to calculate statistical measures (such as mean and standard deviation), and returns the results.

12. Write a program that reads a file of experimental data and detects invalid entries. Generate or use a real dataset. The program should raise exceptions for

common errors (e.g., non-numeric data where numbers are expected) while continuing to process valid data.

13. Modify the lambda function in Section 1.14.4 to filter out measurements that are below a specified error threshold and report the percentage of valid measurements. Add an additional input parameter for the threshold.

14. Extend the car class inheritance example in Section 1.15 to account for fuel and battery capacity constraints for gas and electric cars. 1) Limit their driving distances by checking when the capacities are exceeded. 2) Stop the car when it runs out of fuel or battery power, and print the result.

15. Develop a logging system that writes time-stamped messages to a text file. Include options for errors and warnings. Use file handling and string formatting.

16. Write a class, `ResistorCircuit`, that represents a series or parallel circuit of resistors. Include methods to add resistors, calculate the total resistance, and print the circuit type.

1.21   ADVANCED EXERCISES

1. Recreate the compound data dictionary by calculating molecular weights from another dictionary of element weights. Start by using the symbol strings to deconstruct the contributing elements and their quantities to calculate the compound's molecular weight. For example, SiO_2 contains 1 silicon and 2 oxygen atoms per molecule. Generate additional dictionary entries for compounds used in electronics production or another application. Add other relevant physical properties, and perform operations with the dictionary.

2. Write a program that performs multidimensional interpolation for a given set of input and query points. Apply it to engineering physical properties. For example, it could return interpolated values for steam enthalpy and entropy given input pressure and temperature. Use nested lists to store the table data, and apply linear interpolation for non-tabulated values. Implement bilinear interpolation if necessary for two-dimensional tables.

3. The integration examples covered first-order differential equations. Extend this by developing the capability to solve second-order differential equations and apply it to relevant systems. These systems can be described by a differential equation involving the second derivative of a variable with respect to time. For example, the car class simulation could integrate an acceleration function to compute velocity, and then integrate velocity to determine distance. Other examples include an RLC electrical circuit, a pendulum, a torsional spring with rotational inertia, a vehicle suspension system, a coupled mass system, or a simple harmonic oscillator.

4. Extend Python's built-in `complex` class to support additional operations commonly used in electrical engineering for calculating impedance in AC circuits. Write functions to simulate series and parallel circuits involving complex impedance calculations.

5. Implement a simple finite element analysis (FEA) tool for a 1D beam subjected to a point load. Divide the beam into elements, define stiffness matrices for each element, and solve the system of linear equations to find displacements at each node.

6. Expand upon the `ResistorCircuit` exercise to allow for the design of more complex circuits. Enable arbitrary connections between series and parallel segments of resistors. Write methods to calculate the resistance of each circuit segment and compute the total resistance of the entire circuit.

7. Write a function or class that implements the Newton-Raphson method for solving nonlinear equations. It should accept a mathematical expression, its derivative, and an initial guess as input. Apply this method to solve a nonlinear engineering problem, ensuring the function handles edge cases such as non-convergence or division by zero.

2 General Purpose Scientific and Utility Libraries

The major libraries for general purpose scientific and engineering usage are introduced in this chapter. The big four are *NumPy* as a numerical computing basis, *Pandas* for data analysis, *SciPy* for advanced scientific functions, and *Matplotlib* for plotting and data visualization. They are often used together in various combinations depending on the specific task. They provide generic capabilities for all areas of engineering without specific domain functionality (the only exception is that SciPy has a module for signal processing).

These libraries serve as components or building blocks for other open-source libraries in the larger ecosystem. NumPy is the primary example being the foundation upstream library depended upon by others. This can help simplify integration and data exchange with other Python libraries since they are designed to be interoperable. For example, NumPy arrays and Pandas DataFrames can be used for input in lieu of lists.

The libraries first need to be installed into a working Python development environment. They are normally already installed in modern tools (if not use the `pip` ⤶ `install` or `conda install` commands).

It can be helpful to understand library dependencies. Pandas uses NumPy extensively for its underlying data structures and operations. Its DataFrame and Series objects are built on NumPy arrays. It also uses SciPy and Matplotlib for some functions, but they are not core dependencies. SciPy builds on top of NumPy and extends it with additional scientific functions. Matplotlib depends on NumPy for numerical operations. It also uses SciPy for some functions, but it's not a required library.

All these open-source libraries are free to use, distribute, and modify, making them accessible to everyone. They are designed to be reusable and can be customized to fit specific needs and requirements. They are supported by large, active communities. Code is subject to intense peer review and scrutiny before being put into a public baseline.

Open-source development is at the forefront of innovation, as the libraries are maintained by large communities of developers who are passionate about advancing the state of the art. Users can benefit from ongoing support, documentation, and bug fixes. It is even better to help and participate in the communities.

2.1 NUMPY

NumPy (for **Num**erical **Py**thon) is the de-facto library for scientific computing with powerful tools for multi-dimensional arrays and matrices as essential data structures for engineering applications [14]. NumPy arrays resemble standard Python lists and are used for similar operations, but NumPy is optimized for fast performance on large datasets involving complex calculations. It is written in C and compiled allowing for faster computation.

Beyond its core functionalities in handling arrays and matrices, it has a vast collection of mathematical functions to operate on arrays, capabilities in probability and statistics, and other advanced features. Importantly, NumPy is a basis for other scientific libraries in which it is used internally including SciPy, Pandas, and Matplotlib, and it integrates with many others.

A major advantage of using NumPy is that array vector operations are generally much faster than iterating through lists. Vectorized operations apply functions to entire arrays at once, rather than element by element. This allows for efficient parallel processing. Operations incur execution overhead only once for an array operation compared to the interpretation overhead for each iteration. Furthermore, NumPy arrays are stored in contiguous blocks of memory for faster memory access and better cache utilization.

NumPy's primary object is the multidimensional array as a table of elements, all of the same type, indexed by a tuple of non-negative integers. The dimensions of an array are called axes. A summary of common tasks using NumPy is in Table 2.1. A small sample of basic tasks follow, with dozens more in subsequent chapters.

2.1.1   ARRAY CREATION

After importing NumPy, arrays can be created from Python lists or tuples using the `array` function. The elements of a multi-dimensional array must have the same size.

```python
import numpy as np

# 1D arrays of equidistant load distributions along beam (N/m)
beam_load_external = np.array([0, 80, 160, 240])
beam_load_internal = np.array(227.5, 227.5, 227.5, 227.5)

# 2D array of pressures on a square surface (Pascals)
pressures = np.array([[100., 200.], [300., 400.]])
```

Arrays can be indexed and sliced similarly to Python lists with additional options. Use standard Python syntax for the elements starting at number zero.

```python
>>>pressures[0, 1]
200.0
```

Array attributes include `ndim` (the number of axes), `shape` (the size of each axis), and `dtype` (the data type of the array).

```python
print(beam_load_external.ndim, beam_load_external.shape, ↙
    beam_load_external.dtype)
print(pressures.ndim, pressures.shape, pressures.dtype)
1 (4,) int64
2 (2, 2) float64
```

Table 2.1

Common NumPy Operations

Task	NumPy Functions and Operations
Array Creation and Attributes	
Create from list	`arr = np.array([1, 2, 3])`
Set data type	`arr = np.array([12, 23, 43], dtype=float)`
Access elements with slicing	`arr[0], arr[1:3]`
Get array shape	`arr.shape`
Change data type	`arr.astype(int)`
Operations	
Element-wise operations	`arr1 + arr2, arr * 2`
Mathematical functions	`np.sin(arr), np.sqrt(arr), etc.`
Linear algebra	`np.dot(arr1, arr2), np.linalg.inv(arr)`
Array Manipulation	
Reshape arrays	`arr.reshape(2, 3)`
Concatenate arrays	`np.concatenate([arr1, arr2])`
Split arrays	`np.split(arr, 2)`
Copy arrays	`arr_copy = arr.copy()`
Statistical Functions	
Mean, Median, etc.	`np.mean(arr), np.median(arr)`
Minimum, Maximum	`np.min(arr), np.max(arr)`
Standard deviation	`np.std(arr)`
Random Numbers	
Generate random numbers	`np.random.rand(2, 3)` (uniform)
Generate random integers	`np.random.randint(1, 10, size=5)`
File Input and Output	
Save array	`np.save('data.npy', arr)`
Load array	`loaded_arr = np.load('data.npy')`
Linear Algebra	
Matrix multiplication	`np.dot(arr1, arr2)`
Solving equations	`np.linalg.solve(A, b)`
Eigenvalues and eigenvectors	`np.linalg.eig(arr)`

The shape of arrays can be changed. The following creates an array of 9 values using the `arange` function and re-arranges it into a 3 x 3 matrix using `reshape`.

```
g = np.arange(9).reshape(3, 3) # Create a 3x3 matrix
g
[[0 1 2]
 [3 4 5]
 [6 7 8]]
```

Arrays can also be populated in many others ways such as with zeros to initialize them with random values for probability distributions, or with the frequently used `linspace` function to generate an array at evenly spaced values.

2.1.2 ARRAY OPERATIONS

NumPy arrays support a variety of operations that are performed element-wise faster than standard Python. Basic arithmetic operations like addition, subtraction, multiplication, and division can be performed on arrays.

```
# Array addition
beam_load_total = beam_load_internal + beam_load_external
print(total_load)
[227.5 307.5 387.5 467.5]
```

```
# Array multiplication
# Areas of square surface grid cells (square meters)
areas = np.array([[0.5, 0.5],
                  [0.5, 0.5]])
# Calculate force on each surface element
forces = pressures * areas
print(forces)
[[ 50. 100.]
 [150. 200.]]
```

Multiplication of different shaped arrays are possible. The next operation demonstrates NumPy *broadcasting* to perform arithmetic operations on arrays of different shapes.

```
# Criteria weights
weights = np.array([3, 10, 4])
criteria_performance = np.array([[86.4, 89.2, 75.0],
                 [90.2, 89.0, 82.1 ]])
# Calculate weighted performance for each criteria
weighted_performance = weights * criteria_performance
print(weighted_performance)
[[259.2 892.  300. ]
 [270.6 890.  328.4]]
```

A vectorized operation is performed below converting a NumPy array of Fahrenheit temperatures to Celsius. The expression operates element-wise on the `temperatures_fahrenheit` array, resulting in a new NumPy array containing the Celsius temperatures.

```python
# Fahrenheit temperatures in array
temperatures_fahrenheit = np.array([62, 59, 86, 81, 67, 69])

# Apply vectorized operation
temperatures_celsius = (temperatures_fahrenheit - 32) * 5 / 9

print("Fahrenheit:", temperatures_fahrenheit)
print("Celsius:", temperatures_celsius)
Fahrenheit: [62 59 86 81 67 69]
Celsius: [16.66666667 15.   30   27.22222222 19.44444444 20.55555556]
```

2.1.3   ARRAY FUNCTIONS

NumPy provides numerous mathematical functions that can operate on arrays including statistical functions. It can generate random probability distributions, produce histograms, and perform data binning and various array transformations.

The next example uses the `cumsum` function for a beam load analysis. A triangular load distribution is defined across the beam, the shear force is found as the load integral across the beam using `cumsum`, and finally the bending moment is computed similarly as the integral of shear force. A plot of these arrays against distance would be a natural next step using Matplotlib to visualize them.

```python
import matplotlib.pyplot as plt
import numpy as np

# Discretize a beam
L = 15 # Beam length (m)
n_segments = 10 # Number of discrete segments
x = np.linspace(0, L, n_segments)

# Calculate triangular load distribution
w_max = 1350
w = (w_max * x) / L

# Calculate shear force using cumulative sum
shear_force = np.cumsum(w)

# Calculate bending moment using cumulative sum
bending_moment = np.cumsum(shear_force)

print('load\n', w, '\nshear force\n', shear_force, '\nbending ⤸
    moment\n', bending_moment)
```

```
load
[   0.  150.  300.  450.  600.  750.  900. 1050. 1200. 1350.]
shear force
[   0.  150.  450.  900. 1500. 2250. 3150. 4200. 5400. 6750.]
bending moment
[    0.   150.   600.  1500.  3000.  5250.  8400. 12600. 18000. 24750.]
```

The next example creates arrays for projectile time and distance. It includes NumPy trigonometric functions, the `linspace` function to create an array of evenly spaced discrete time points, and generates a distance array of the same length created by multiplication. These arrays will be used later for plotting trajectories vs. angle.

```python
# create projectile time and distance arrays

import numpy as np

def flight_time(v0, angle_deg, g=9.81):
    angle_rad = np.radians(angle_deg)
    return 2 * v0* np.sin(angle_rad) / g

v0 = 100 # initial velocity (m/s)
angle = 45 # launch angle (degrees)

time_of_flight = flight_time(v0, angle)
time_points = np.linspace(0, time_of_flight, num=10) # time point array
distances = v0* np.cos(np.radians(angle)) * time_points # distance ⤸
    array

print(f'{time_points=} \n {distances=}')
time_points=array([ 0.   ,  1.60178227,  3.20356453,  4.8053468 ,  6.40712906,
        8.00891133,  9.61069359, 11.21247586, 12.81425813, 14.41604039])
 distances=array([ 0.   ,  113.26311021,  226.52622041,  339.78933062,
        453.05244082,  566.31555103,  679.57866123,  792.84177144,
        906.10488164, 1019.36799185])
```

There are functions for linear algebra operations like matrix multiplication, eigenvalues, and solving linear systems. NumPy can be used for solving systems of linear equations. The following equations determine the concentrations of three coupled reactors (c_1, c_2, c_3) as a function of the amount of mass input in g/day to each reactor shown on the right sides of the equations.

$$17c_1 - 2c_2 - 3c_3 = 500$$
$$-5c_1 + 21c_2 - 2c_3 = 200$$
$$-5c_1 - 5c_2 - 22c_3 = 30$$

Using the `linalg.solve` function, the coefficients and constants are represented in arrays. The solution provides the values of c_1, c_2, c_3 that satisfy the system of equations.

```
A = np.array([[17, -2, -3], [-5, 21, -2], [-5, -5, 22]])
B = np.array([500, 200, 30])
solution = np.linalg.solve(A, B)
print(solution)
```

```
[33.99631415, 18.89282676, 13.38389566]
```

The full set of available mathematical functions is found at https://numpy.org/
doc/stable/reference/routines.math.html.

2.1.4 STATISTICS AND PROBABILITY DISTRIBUTIONS

NumPy has basic statistical functions like mean, median, var, np.std, and more
for summarizing and understanding data distributions. These can be applied to both
NumPy arrays and standard lists.

Aggregation functions like sum, prod, cumsum, and cumprod are helpful in
computing cumulative and multiplicative statistics on arrays. Functions such as
np.corrcoef and np.cov are available for calculating correlation and covariance
matrices to help understand relationships between variables.

The random module includes methods to generate random numbers from probability distributions which are useful in simulations and random sampling. The following creates an array of 1000 values of a Rayleigh distribution to model wind velocity.
The scale is the mean value, and the size parameter is the number of samples. This
method of generating random samples is used to drive a simulation of wind turbine
power output in Section 3.13.

```
np.random.rayleigh(scale=5, size=1000)
array([ 8.54395347,  6.29693514,  6.81646363,  6.59769644,  6.484059 ,
        0.79192554,  3.91892555,  2.59195087,  9.20255815,  3.26062015,
        1.04682111, 10.16864214,  6.76128808,  4.40085928,  8.34416138,
        ...
```

NumPy can compute histograms and bin data for statistical analysis, visualization, and understanding of data distributions. Matplotlib can do much of the same,
though NumPy improves upon it. For example, continuous cumulative distributions
can be computed and custom bins specified precisely with NumPy compared to Matplotlib histogram defaults. See Section 2.2.

Data transformation functions for sorting, filtering, and applying Boolean conditions are available. The following sorts testing data to generate a cumulative probability distribution. The arange function returns evenly spaced values within a given
interval. This example continues in Section 3.1.3 to plot the distribution and use it
for an inverse transform method to generate random variates.

```
# testing duration times
durations = [8.9, 4.6, 1.6, 3.5, 11.8, 1.5, 6.0, 7.0, 1.8, 1.9, 2.8, ↙
    1.9]

# sort data
```

```
sorted_durations = np.sort(durations)

# create cumulative count array of evenly spaced values length of ↵
    the sorted data [0, 1, 2, ...] and normalize it 0-1
cum_probabilities = ↵
    np.arange(len(sorted_durations))/float(len(sorted_durations)-1)
print(cum_probabilities)
[0.         0.09090909 0.18181818 0.27272727 0.36363636 0.45454545
 0.54545455 0.63636364 0.72727273 0.81818182 0.90909091 1.        ]
```

2.2 MATPLOTLIB

Matplotlib is a fundamental library for creating static, animated, and interactive visualizations. It is widely used for data analysis and exploration in engineering and science, providing many customization options for creating publication-quality graphics. These include line plots, scatter plots, bar plots, histograms, continuous distributions, others, and combinations. A gallery of examples is at [12].

Plots can be specified with data input consisting of pairwise data, sequences for distributions, grid data and 3D and volumetric data. It can generate different graphic formats for vector or bitmapped static images, and animated videos on a range of output backends. Matplotlib is integrated with NumPy, Pandas, and SciPy. Conversely it is used by other libraries to produce graphics.

Matplotlib produces displays in virtually all Python environments except basic interpreters with only a command line interface. It has an active development community, and is well documented with numerous examples online.

2.2.1 PLOTTING FUNDAMENTALS

Central to Matplotlib is *Pyplot*, a comprehensive collection of functions for adding or customizing plot elements such as lines, axes, ticks, images, text, labels, and other decorations. These are identified in Figure 2.1, showing the anatomy of a figure as components and methods to manipulate them for a combined line and scatter plot.

As illustrated, a *Figure* is the larger canvas that *Axes* are placed within. Each pyplot function makes some change to a figure, such as create a figure, create a set of axes within a figure, plot some lines, decorate the plot with labels, etc. Examples in Figure 2.1 show lines are drawn with `ax.plot` and scatter plots with `ax.scatter`. The axis labels can be set with the `ax.set_xlabel` and `ax.set_ylabel` methods respectively, and other components similarly as shown.

Matplotlib has both an object-based *Axes* interface and a function-based *pyplot* interface. The *Axes* interface is used to create a figure and one or more axes objects; then methods are used explicitly on the named objects to add data, configure limits, set labels etc. The *pyplot* interface consists of functions to manipulate figures and axes that are implicitly present (also called state-based). *pyplot* is mainly intended for rapid plots and simple cases of plot generation. Many functions are identical in both interfaces though the object-oriented API provides more options and flexibility.

Table 2.2 contrasts the two interfaces to draw an identical line plot. The essential Matplotlib module is typically imported with `import matplotlib.pyplot as plt`. The primary statement in both is the identical `plot` method. Note that the end statement `show()` isn't necessary in many newer environments to display the plots. A `plt.savefig()` function would be used to save a graphic file.

An extra statement is needed with the *Axes* interface to instantiate figure and axis objects, but the *pyplot* approach has limitations and will require more bookkeeping overhead with additional plots. Complex and detailed plots are usually simpler with the explicit O-O interface compared to using the implicit *pyplot* interface.

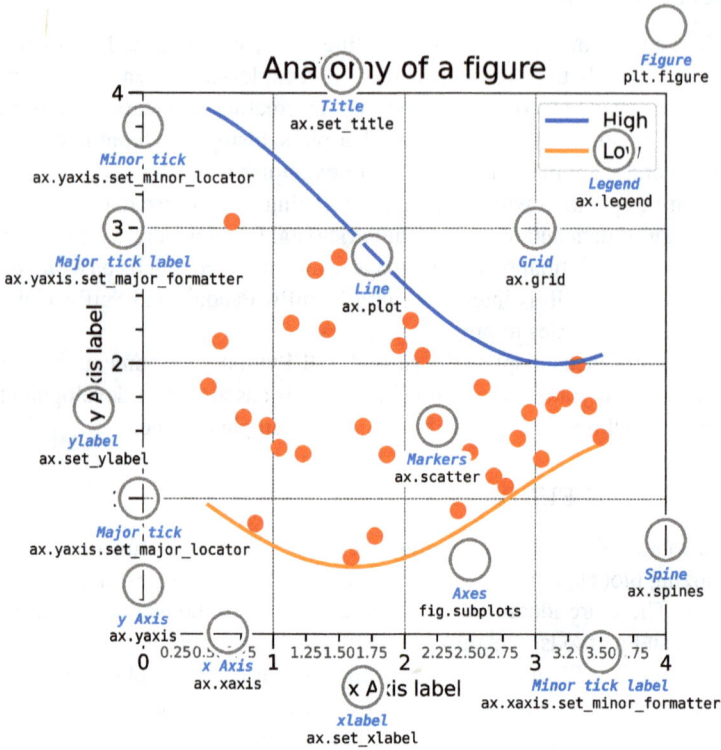

Figure 2.1   Matplotlib Figure Component Anatomy and Methods

Table 2.2
Matplotlib Interfaces

Axes interface	pyplot interface

`import matplotlib.pyplot as plt`
 	
`fig, ax = plt.subplots()`
`ax.plot(x, y)`
`ax.set_title("Engineering Data")`
`plt.show()`
	`import matplotlib.pyplot as plt`
	
	`plt.plot(x, y)`
	`plt.title("Engineering Data")`
	`plt.show()`

Table 2.3 shows common Matplotlib operations. These examples use the Axes interface and assume the figure and axis object are already created with `fig, ax` ⤸ `= plt.subplots()`. The plot type functions work the same with the pyplot interface

using the `plt` name instead of `ax`. For more examples, a set of Matplotlib cheatsheets for beginning to advanced users is available on GitHub at https://github.com/matplotlib/cheatsheets. They contain numerous options and detailed reference beyond this introduction.

Table 2.3: Common Matplotlib Operations

Task	Examples
Line and Scatter Plots	
Line plot with X and Y	`ax.plot(x, y)`
Plot with markers and line style	`ax.plot(x, y, 'o--')`
Multiple line plots	`ax.plot(x1, y1, x2, y2)`
Customize plot attributes	`ax.plot(x, y, color='green', label='Data')`
Plot dictionary data	`{data = 'x': x, 'y': y}`
	`ax.plot('x', 'y', data=data)`
Scatter plot	`ax.scatter(x, y)`
Scatter plot with color and size	`ax.scatter(x, y, c='blue', s=50)`
Histogram	
Histogram	`ax.hist(data)`
Histogram with bins and density	`ax.hist(data, bins=20, density=True)`
Cumulative histogram	`ax.hist(data, bins=20, cumulative=True)`
Bar Plot	
Bar plot	`ax.bar(x, height)`
Horizontal bar plot	`ax.barh(x, height)`
Bar plot with custom colors	`ax.bar(x, height, color='orange')`
Box Plot	
Box plot	`ax.boxplot(data)`
Box plot with custom whiskers	`ax.boxplot(data, whis=[5, 95])`
Pie Chart	
Pie chart	`ax.pie(sizes, labels=labels)`
Pie chart with custom colors and explode	`ax.pie(sizes, labels=labels, ⤶ colors=colors, explode=explode)`
3D Plot	
3D scatter plot	`ax.scatter3D(x, y, z)`
3D line plot	`ax.plot3D(x, y, z)`

Continued on next page

Table 2.3: Common Matplotlib Operations – Continued

Task	Examples
Other Plot Types	
Error bars plot	`ax.errorbar(x, y, yerr=error)`
Heatmap	`cax = ax.imshow(data, cmap='hot')`
Plot Customization	
Add title	`ax.set_title('Title')`
Set X and Y labels	`ax.set_xlabel('X-axis')`
	`ax.set_ylabel('Y-axis')`
Set limits	`ax.set_xlim([xmin, xmax])`,
	`ax.set_ylim([ymin, ymax])`
Add legend	`ax.legend(loc='upper left')`
Set multiple attributes	`ax.set(title='Title', xlabel='X-axis',` ⤶
	`ylabel='Y-axis')`
Saving and Showing Plots	
Show plot	`plt.show()`
Save plot to file	`fig.savefig('plot.png')`
Matplotlib Defaults Customization	
Change global default settings	`plt.rcParams['font.size'] = 14`
Use a custom style	`plt.style.use('seaborn-darkgrid')`
Temporarily set defaults with `plt.rc_context()`	`with plt.rc_context({'lines.linewidth':` ⤶
	`2.5}):`
	`    ax.plot(x, y)`
Use `matplotlib.rc` for permanent settings	`matplotlib.rc('lines', linewidth=2,` ⤶
	`color='r')`
Restore default settings	`plt.rcdefaults()`

2.2.2   PLOT FUNCTION

The basic `plot` function is versatile and will take an arbitrary number of arguments per the call signature:

```
plot([x], y, [fmt], *, data=None, **kwargs)
```

The x and y data sequences are the coordinates for lines or points. The optional parameter *fmt* specifies basic formatting like color, marker, and linestyle (e.g., `'r'` or `'red'` specifies red color). The optional *data* parameter is an object with labelled data in lieu of x and y sequences (e.g., a dictionary or Pandas DataFrame), and the labels are provided for plotting. Optional keyword arguments (*kwargs*) provided as a dictionary will be unpacked to specify plot properties such as a line label, linewidth, marker face colors, etc. These inputs are shown in successive examples.

2.2.3 LINE AND SCATTER PLOTS

Matplotlib can plot y versus x as lines and/or unconnected markers for scatter plots. The minimum input is a set of y values. When provided a single sequence, it assumes y values and automatically generates an x sequence starting with 0 of the same length. In this example, a list of 24 measurements is sent to generate the line plot in Figure 2.2. After importing pyplot into the conventional namespace `plt`, only `plt.plot()` is necessary to draw a line. Labels for the axes are also added.

```python
import matplotlib.pyplot as plt

temperatures = [303, 341, 315, 320, 301, 320, 330, 330, 323, 309, 310
    , 330, 333, 320, 310, 330, 323, 299, 310, 309, 293, 300, 310, 314]

# input y values
plt.plot(temperatures)
plt.xlabel('Hour')
plt.ylabel('Tank Temperature (K)')
```

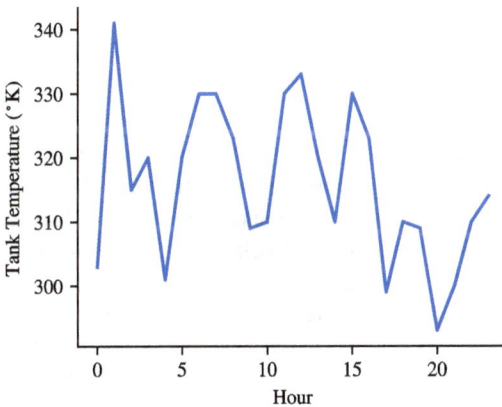

Figure 2.2 Line Plot Given Y Values

The presumption of x values as integers starting with zero corresponds to actual hour values here, which is rarely the case. The x-axis tick labels could be modified or both x and y values provided per subsequent examples.

The next two lists are input for both x and y values to plot height vs. velocity for the projectile launcher in Figure 2.3.

```python
import matplotlib.pyplot as plt
import numpy as np

g = 9.81 # m/s^2
velocities = np.linspace(20, 100, 9)
heights = [velocity**2/g/2 for velocity in velocities]

# input x and y data lists
plt.plot(velocities, heights)
plt.xlabel('Velocity (m/s)')
plt.ylabel('Height (m)')
```

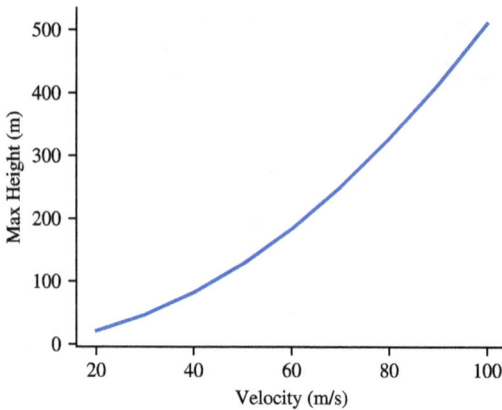

Figure 2.3   Line Plot Given X and Y Values

The *pyplot* API used in the previous examples is rapid but less flexible than the object-oriented *Axes* interface primarily used in this book. When generating multiple plots in a script, *pyplot* becomes trickier and more verbose. The plot state needs re-initializing between plots with a `clf()` to clear the current figure or a `cla()` to clear the current axes.

With the *Axes* interface, Figure and Axes objects are created using `plt.subplots()`. The `subplots` method returns figure and axis objects that are typically assigned to `fig` and `ax` respectively. Object methods are then used to draw data such as `ax.plot()` or `ax.scatter()` and set properties such as `ax.set_xlabel()` or `ax.set()` which can take multiple attribute settings.

The *Axes* interface is used to plot multiple flight time curves against varying launch angles in Figure 2.4. A dictionary is created with trajectory data for a specified parameter space and plotted. The dictionary contains lists of flight times for each angle keyed by velocity values. Each `ax.plot` in the loop adds another velocity curve.

```python
import matplotlib.pyplot as plt
import math
import numpy as np

def flight_time(v0, angle):
    return(2 * v0* math.sin(0.01745 * angle) / 9.8)

# Parameter space
angles = np.linspace(10, 90, 10)
velocities = np.linspace(50, 80, 4)

# Populate trajectory data dictionary
trajectory_data = {velocity: [flight_time(velocity, angle) for angle ∠
    in angles] for velocity in velocities}

# Plot flight times curves against angle for each velocity
fig, ax = plt.subplots()
for velocity in velocities:
    ax.plot(angles, trajectory_data[velocity], label=f'v0 = ∠
        {velocity}')
ax.set(xlabel = 'Launch Angle (degrees)', ylabel="Flight Time ∠
    (seconds)")
ax.legend()
```

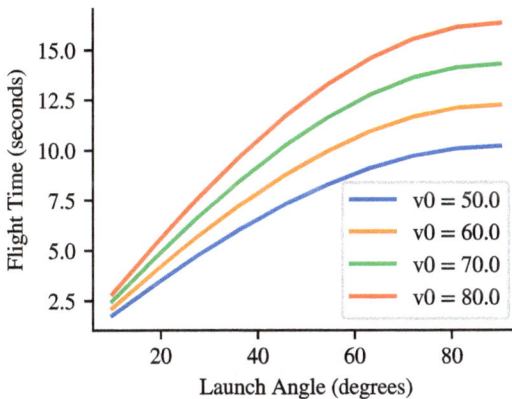

Figure 2.4   Multiple Line Plot

A scatter plot can be drawn with the same `plot` command used for a line by spec-ifying a marker type (e.g., with a shorthand parameter `'o'` or `marker='x'`) or with the `scatter` method. The next example uses a scatter plot to visualize and characterize cost functions for different types of engines from historical data. Each engine type is plotted as a separate call and color coded. The NumPy line fitting function `polyfit` is used to derive a best fit regression line and plotted for each engine type.

   This example reads input from a CSV file with the following structure. Note the
manual file processing and data analysis could be simpler using a general purpose
Pandas function per the next section.

```
cost,horsepower,engine_type
21343.6,326,diesel
34990.2,658,diesel
54330.3,586,gas_turbine
...
```

```python
import matplotlib.pyplot as plt
import numpy as np

# Read CSV file and populate dictionary.
filename = 'engine cost and horsepower data.csv'
engine_data = {}
with open(filename, 'r') as file:
    next(file) # Skip header line
    for line in file:
        cost, horsepower, engine_type = line.strip().split(',')
        cost = float(cost)
        horsepower = float(horsepower)

        if engine_type not in engine_data:
            engine_data[engine_type] = {'cost': [], 'horsepower': []}
        engine_data[engine_type]['cost'].append(cost)
        engine_data[engine_type]['horsepower'].append(horsepower)

# Scatter plots and regression lines
fig, ax = plt.subplots()
colors = {'electric': 'green', 'diesel': 'red', 'gas_turbine': 'blue'}
for engine_type, values in engine_data.items():
    x = np.array(values['horsepower'])
    y = np.array(values['cost'])

    # Fit a linear regression line and get coefficients
    slope, intercept = np.polyfit(x, y, 1)

    # Create x and y values for the line
    x_line = np.linspace(min(x), max(x), 2)
    y_line = slope * x_line + intercept

    plt.scatter(x, y, s=7, color=colors[engine_type], ↵
        label=engine_type)
    plt.plot(x_line, y_line, color=colors[engine_type], linestyle='--')

ax.set(xlabel='Horsepower', ylabel='Development Cost (k)')
ax.legend()
```

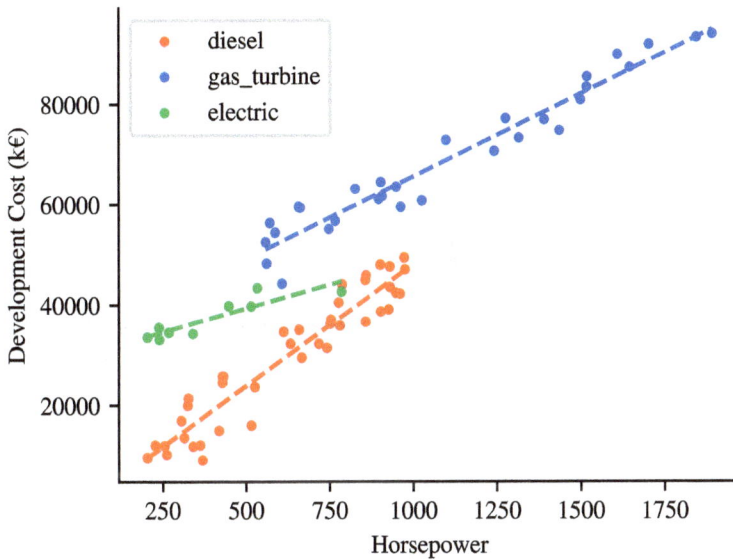

Figure 2.5   Scatter and Line Plot

2.2.4   HISTOGRAMS AND DISTRIBUTIONS

The necessary input to generate a histogram or other probability distribution is a data sequence in a list, NumPy array, or labeled data. In this example, random wind velocities for a wind turbine simulation are modeled as a lognormal distribution and displayed as a frequency histogram in Figure 2.6.

```python
import numpy as np
import matplotlib.pyplot as plt

# Generate 5000 random wind velocity values using lognormal ↲
    distribution
# Mean velocity of underlying normal distribution (mph)
wind_mean = 3.5
# Standard deviation of underlying normal distribution (mph)
wind_std = .56
velocities = np.random.lognormal(wind_mean, wind_std, 5000)

# Histogram
fig, axis = plt.subplots()
axis.hist(velocities, bins=100)
axis.set(xlabel = 'Wind Velocity', ylabel='Frequency', xlim = [0, 200])
```

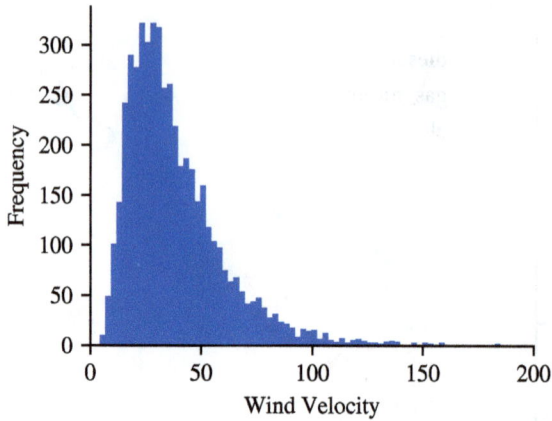

Figure 2.6   Histogram

The cumulative form of a distribution can be specified with the optional parameter `cumulative` as shown next. The option `histtype='step'` produces a line plot without being filled per Figure 2.7.

```
# Cumulative distribution
fig, axis = plt.subplots()
axis.hist(velocities, histtype='step', cumulative=True, bins=100)
axis.set(xlabel = 'Wind Velocity', ylabel='Cumulative ↙
    Probability',xlim = [0, 200])
```

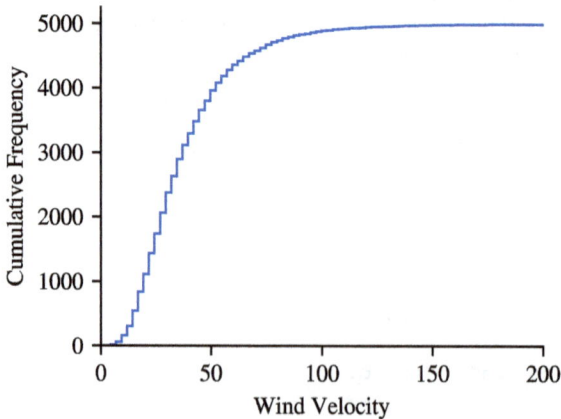

Figure 2.7   Cumulative Histogram

2.2.5 BOX PLOTS

Box plots are also created from data sequences in lists, arrays, or labeled series. In this example the targeting errors for different types of projectile launchers are modeled as normal distributions. Both arrays are sent and displayed alongside each other for easy comparison as seen in Figure 2.8.

```python
import matplotlib.pyplot as plt

# Targeting error normal distributions for 300 meter target
target_error_mean = 0 # mean error (m)
target_error_auto_std = 4 # standard deviation (m)
target_error_manual_std = 9 # standard deviation (m)

# Generate 1000 random error values using normal distribution
errors_auto = np.random.normal(target_error_mean, ↙
    target_error_auto_std, 1000)
errors_manual = np.random.normal(target_error_mean, ↙
    target_error_manual_std, 1000)

fig, ax = plt.subplots()
ax.boxplot([errors_auto, errors_manual])
ax.set_xticklabels(['Electronic', 'Manual'])
ax.set(xlabel = "Launcher Type", ylabel="Targeting Error (m)")
```

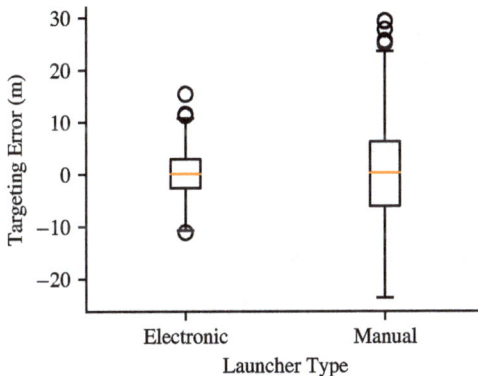

Figure 2.8   Boxplot

2.2.6   3-DIMENSIONAL PLOTS

Three dimensional plots require input arrays for the X, Y, and Z axes. Matplotlib can visualize either surface or volumetric data in 3D space, as well as lines, histograms, and other plot types. They can also be animated (see the next section). Many examples of 3D plotting can be found at https://matplotlib.org/stable/gallery/mplot3d/index.html. Color maps for 3D plotting are described in detail at https://matplotlib.org/stable/users/explain/colors/colormaps.html.

In the next example, a response surface visualization of projectile distance by angle and velocity is created with the `plot_surface` method. A color map is specified with the `cmap` parameter to visualize the response surface in Figure 2.9.

```python
from math import sin, cos
import numpy as np
import matplotlib.pyplot as plt

def projectile(v0, angle):
    """ Returns the projectile flight time, maximum height and ↵
        distance given initial velocity in meters per second and ↵
        launch angle in degrees. """

    g = 9.8# gravity (meters per second squared)
    angle_radians = 0.01745 * angle # convert degrees to radians
    flight_time = 2 * v0* sin(angle_radians) / g
    max_height = (1 / (2 * g)) * (v0 * math.sin(0.01745 * angle)) ** ↵
        2
    distance = 2 * v0** 2 / g * sin(angle_radians) * cos(angle_radians)
    return(flight_time, max_height, distance)

# Define ranges for velocity and angle
velocity_range = np.arange(0, 100, 10)
angle_range = np.arange(0, 90, 15)

# Create meshgrid for velocity and angle
velocity, angle = np.meshgrid(velocity_range, angle_range)

# Calculate distance for each combination of velocity and angle ↵
    using list comprehension
distances = np.array([[projectile(v, a)[2] for v in velocity_range] ↵
    for a in angle_range])

# Plotting
fig, ax = plt.subplots(subplot_kw={'projection': '3d'})
ax.plot_surface(velocity, angle, distances, cmap='viridis')
ax.set(xlabel = 'Velocity (m/s)', ylabel ='Angle (degrees)', zlabel ↵
    = 'Distance (m)', ) # 'Response Surface of Distance of a ↵
    Projectile'
ax.view_init(elev=20, azim=35, roll=0)
```

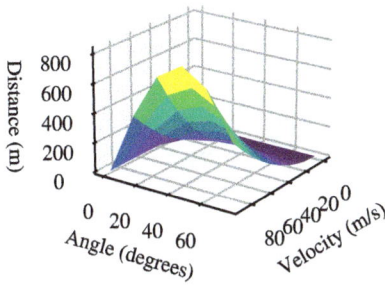

Figure 2.9   Response Surface

2.2.7   ANIMATION

A simple animation can be generated manually in a time loop with a `pause` function between successive draws. The following displays a projectile trajectory as a dynamic scatter plot per the final view in Figure 2.10. With additional elements, this approach may start to lag and require optimization.

```
...
fig, ax = plt.subplots()
ax.set(xlabel='Distance (m)', ylabel='Height (m)', xlim=[0, ⬐
    x_max+1], ylim=[0, y_max+1], title='Projectile Animation')

# Draw projectile moving over time as scatter plot points
for time in np.arange(0, flight_time, dt):
  x, y = projectile_position(time)
  ax.scatter(x, y, marker='o', color='b')
  plt.draw()
  plt.pause(0.1 ) # animation time delay between points (seconds)
```

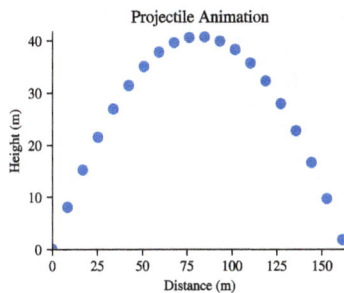

Figure 2.10   Simple Animation of Projectile Trajectory

Matplotlib has general purpose classes for complex optimized animations and can save them as video files. It offers the primary methods `FuncAnimation` and `ArtistAnimation` for creating animations:

- `FuncAnimation` creates animations by repeatedly calling a function to update the plot. A function needs to be explicitly defined that computes the updates for each frame. This method is appropriate when the plot data changes over time requiring calculations. The basic template for a repeating animation with `FuncAnimation` is in Listing 2.1.

- `ArtistAnimation` creates animations by using fixed sets of *Artist* objects (lines, patches, etc.). It directly manipulates the artists and passes a sequence of them to the animation function. `ArtistAnimation` is best for animations with pre-computed data to be displayed sequentially, or with complex visual elements that are difficult to update with `FuncAnimation`.

Listing 2.1 Animation Template

```python
import matplotlib.pyplot as plt
import matplotlib.animation as animation

# Create a figure and axis
fig, ax = plt.subplots()

# Initialization function for each animation repeat
def init():
    # Draw clear frame and initialize data

# Animation update function for each frame
def animate(frame):
    # Plot new data

# Create the animation
anim = animation.FuncAnimation(fig, animate, init_func=init, ⤸
    frames=10, interval=500, repeat=True)

# Show the animation
plt.show()
```

A full implementation of a projectile trajectory animation for a fixed case is demonstrated next in Listing 2.11 and shown mid-animation in Figure 2.11. This approach draws a trajectory history line, displays a moving circle for the projectile over time, and writes the ongoing state as the simulation progresses. An init function would be used if the launch parameters are to change for each animation repeat.

For smoother animations which may become necessary with added complexity, use `blit=True` in both FuncAnimation and ArtistAnimation. The initialization and update routines must then explicitly return all objects being updated. Other animation parameters like `repeat` and `save_count` can be used to customize animations. Use `matplotlib.widgets` to create user interfaces for interactive animations.

Listing 2.2 Projectile Animation

```python
import numpy as np
import matplotlib.pyplot as plt
import matplotlib.animation as animation

# Parameters
v0 = 60
theta = np.pi / 4 # Launch angle (radians)
g = 9.81

# Calculate trajectory over time
flight_time = 2 * v0* np.sin(theta) / g
t = np.linspace(0, flight_time, 50) # 50 time points
y = v0* np.sin(theta) * t - 0.5* g * t**2
x = v0* np.cos(theta) * t

# Plot
fig, ax = plt.subplots()
trajectory, = ax.plot([], [])
projectile, = ax.plot([], [], marker='o', color='red')
display_text = ax.set_title('')
ax.set(xlim=[0, max(x)], ylim=[0, max(y)], xlabel='Distance (m)', ↙
    ylabel='Height (m)')

# Update the plot
def update(frame):
    # Moving projectile shown at current time
    projectile.set_data(x[frame], y[frame])
    # Trajectory line displayed up to current time
    trajectory.set_data(x[:frame], y[:frame])
    # Display state over time
    display_text.set_text(f'Time: {t[frame]:.2f} s \nDistance: ↙
        {x[frame]:.2f} m \nHeight: {y[frame]:.2f} m')

anim = animation.FuncAnimation(fig=fig, func=update, frames=len(t), ↙
    interval=30)
plt.show()
```

Note that the animation API call method mirrors the logic of a general purpose time-based simulation. As described in [11], a simulation framework requires a time clock, parameter initialization routine, and update procedure for each time step which corresponds to a frame in Matplotlib. Thus the animation template can be easily and naturally integrated with dynamic simulations in various ways (see Chapter exercises). They can also be made interactive with user controls.

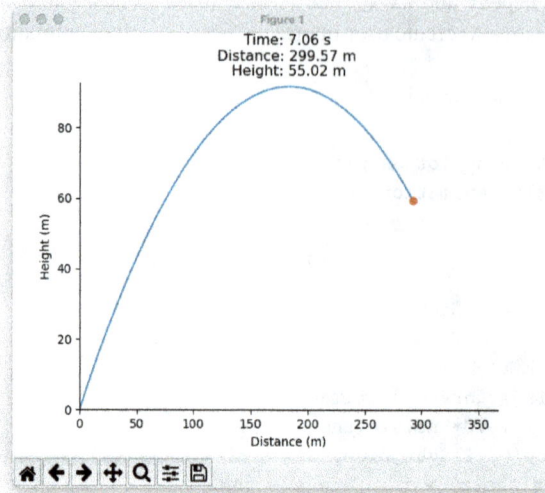

Figure 2.11   Projectile Animation

2.3   PANDAS

Pandas is the *Python Data Analysis Library* used to explore and analyze data in many formats. It provides data structures for efficient data manipulation and rapid analysis with functions for statistics, time series analysis, data cleaning, merging, reshaping and more. It is widely used in engineering applications for large datasets including simulations and machine learning. Being compatible with NumPy, SciPy, Matplotlib, statsmodels, scikit-learn, and many other libraries, it saves effort with its richness of underlying data structures and functions.

Central to Pandas are data structures for a *Series* as a one-dimensional array holding data of any type and a *DataFrame* as a two-dimensional table similar to a spreadsheet. A DataFrame object contains rows and columns with additional labelling metadata. DataFrames are powerful for data manipulation and analysis involving multiple related variables.

Data structures are labeled with row indices and column names. Pandas by default will create its own index for each Series or DataFrame row starting from 0 and incrementing by 1 for each element. Other indices can also be specified. Each DataFrame column is labeled with a column name and acts like a Series.

A summary cheatsheet of common tasks using Pandas is in Table 2.4 showing the breadth of available operations. The reader can find a thorough coverage of using DataFrames for data manipulation for engineering applications in [13]. Examples of basic tasks for creating datasets, selecting and indexing data, generating statistics, and vectorized operations follow.

Table 2.4: Common Pandas Operations

Task	Examples
Data Loading and Saving	
Load CSV	`df = pd.read_csv('data.csv')`
Load Excel (xlsx)	`df = pd.read_excel('data.xlsx')`
Save CSV	`df.to_csv('data.csv', index=False)`
Save Excel (xlsx)	`df.to_excel('data.xlsx', index=False)`
Selection and Indexing	
Select column(s)	`df[['column1', 'column2']]`
Select rows by label	`df.loc[row_label]`
Select rows by condition	`df[df['column'] > value]`
Get column by label	`df['column_name']`
Iteration	
By row	`for index, row in df.iterrows():`
	`for i, row_index in enumerate(df.index):`
By column	`for column in df:`

Continued on next page

Table 2.4: Common Pandas Operations – Continued

Task	Examples
Data Manipulation	
Add new column	`df['new_column'] = ...`
Remove column	`del df['column']`
Add, remove rows	`df.append(...), df.drop(index)`
Rename columns	`df.rename(columns='old_name': 'new_name')`
Sort DataFrame	`df.sort_values(by='column')`
Missing Values	
Check for missing values	`df.isnull().sum()`
Drop rows with missing values	`df.dropna()`
Fill missing values	`df.fillna(value)`
Aggregation and Statistics	
Describe data	`df.describe()`
Statistics	`df['column'].mean()`
	`df['column'].median()`
	`df['column'].std()`
Group-by operations	`df.groupby('column').mean()`
Count unique values	`df['column'].value_counts()`
Time Series	
Set time as index	`df.set_index('time_column')`
Resample data	`df.resample('D').mean()`
Merge and Join	
Concatenate DataFrames	`pd.concat([df1, df2])`
Merge DataFrames on key	`df = pd.merge(df1, df2, on='column_name')`
Join DataFrames on key	`df = df1.join(df2, on='column_name')`
Apply Functions	
Apply function to each element	`df['column'] = ↙`
	`df['column'].apply(function)`
Vectorized operations	Use operators and functions directly on DataFrame or Series
Other Useful Functions	
Show first, last few rows	`.head(), .tail()`
DataFrame information	`.info()`

2.3.1   CREATING DATASETS

In this example a series of pressure measurements is created from a list and assigned to a Pandas Series. A Series or DataFrame is typically named with the `df` abbreviation as done here. The Series is printed showing its internally created indices, an element is accessed with its index, and the statistical mean function is used.

```python
import pandas as pd

# Create data series
pressures = [22, 24, 24, 25, 27, 29, 30, 27, 26, 26] # PSIA
df = pd.Series(pressures)

print(df)

# Access elements by index
print(f"Third pressure reading: {df[2]} PSIA")

# Pressure mean
print(f"Mean pressure: {df.mean()} PSIA")
```

```
0    22
1    24
2    24
3    25
4    27
5    29
6    30
7    27
8    26
9    26
dtype: int64
Third pressure reading: 24 PSIA
Mean pressure: 26.0 PSIA
```

Often a custom index is desired that may be part of the data itself or created independently. Next an index is defined and a series element is accessed with its custom index.

```python
import pandas as pd

# Create a Series with custom index labels
temperature_settings = [21, -45, 14, 18] # °C
chemical_tanks = ["Naphtha", "Liquefied Petroleum Gas", "Benzene", ↙
    "Toluene"]
df = pd.Series(temperature_settings, index=chemical_tanks)

print(df)

# Accessing elements by label
```

```
benzene_setting = df["Benzene"]
print(f"Temperature setting of Benzene: {benzene_setting}°C")
```

```
Naphtha                     21
Liquefied Petroleum Gas    -45
Benzene                     14
Toluene                     18
dtype: int64
Temperature setting of Benzene: 14°C
```

2.3.2   FILE READING

External files can be imported and used for datasets. The next example reads an Excel
file containing detailed altitude and air density values for atmospheric calculations
(this is used in Section 3.16 for meteor trajectory analysis). The air density data
points are measured every 2000 feet per the input spreadsheet shown in Figure 2.12.

This example also demonstrates the use of labels for each column. In this case
they come with the data in a header row, but could also be specified manually.

	A	B
1	Altitude	Density
2	0.0	1.23
3	2000	1.01
4	4000	0.819
5	6000	0.66
6	8000	0.526
7	10000	0.414
8	12000	0.312
9	14000	0.228
101	198000	3.60E-10
102	200000	2.50E-10

Figure 2.12   Altitude Density Spreadsheet

The file contents are put directly into the DataFrame using the `read_excel`
function.

```
import pandas as pd

# Create dataframe
df = pd.read_excel('air densities.xlsx')
print(df)
```

```
      Altitude        Density
0        0.0    1.230000e+00
1        2000   1.010000e+00
2        4000   8.190000e-01
3        6000   6.600000e-01
4        8000   5.260000e-01
..        ...            ...
96      192000   3.600000e-10
97      194000   3.600000e-10
98      196000   3.600000e-10
99      198000   3.600000e-10
100     200000   2.500000e-10

[101 rows x 2 columns]
```

One can then select DataFrame columns with their column names.

```
print(df['Altitude'])
```

```
Altitude
0.0       1.230000e+00
2000      1.010000e+00
                ...
198000    3.600000e-10
200000    2.500000e-10
Name: Density, Length: 101, dtype: float64
```

In this case it is desired to set the index appropriately for altitude dependent calculations. The next example defines the altitude column as the index which is seen in the printed summary. It can now be used to relate density for given altitudes in simulations.

```python
import pandas as pd

# Create dataframe and set index
df = pd.read_excel('air densities.xlsx')
df = df.set_index('Altitude')

# Access density value for given altitude
density = df.loc[16000, 'Density']
print(f'Density at 16000 feet = {density}')
```

```
Density at 16000 feet = 0.166
```

2.3.3 DATAFRAME ITERATION

Pandas offers several methods to iterate through a DataFrame by rows and columns per Table 2.4. A common approach for iterating by rows uses the `.iterrows()` function that yields tuples for each row. The first tuple element is the row index, and the second is a Series containing the row's data. Another way is tor use `enumerate()` with the Pandas `.index` function to iterate over both the index and the data in each row.

A DataFrame is directly iteratable by column with a `for column in df:` loop. This iterates and provides the column names where one can access the corresponding column data using `df[column]`.

Another method uses the `.itertuples()` function which is similar to `.iterrows()` but returns named tuples instead of Series. This can be useful to directly access column names within the loop.

For an example of iteration, a daily log file of tank temperatures records serial measurements every 5 minutes for a set of tanks. It is written to a CSV file with the following structure:

```
time,tank_1,tank_3,tank_4,tank_5
0:00,415,302,245,303
0:05,414,301,216,309
0:10,417,308,225,317
...
23:50,417,308,232,320'56
23:55,417,319,215,319
```

The rows for each data collection time point can be iterated over with `.iterrows()`.

```
df=pd.read_csv('tank_levels.csv', index_col='time')

for index, row in df.iterrows():
    print('index =',index, '\nrow = ', row)
index = 0:00
row =  tank_1    415
tank_3    302
tank_4    245
tank_5    303
Name: 0:00, dtype: object
index = 1:00
row =  tank_1    414
tank_3    301
tank_4    216
tank_5    309
...
```

The columns for each tank can be iterated over and accessed for calculations by iterating directly over the dataframe.

```
for tank in df:
    print('tank =', tank, 'mean = ', df[tank].mean())
```

```
tank = tank_1 mean =   414.8333333333333
tank = tank_3 mean =   307.5416666666667
tank = tank_4 mean =   225.20833333333334
tank = tank_5 mean =   310.375
```

2.3.4 APPLYING FUNCTIONS AND VECTORIZED OPERATIONS

The previous section showed that iterating through DataFrames is possible, but it's generally less efficient than applying custom functions or vectorized operations across an entire DataFrame at once. Table 2.4 shows one can apply functions using either the Pandas `.apply()` for custom functions or vectorized operations using Pandas or NumPy built-in functions and methods directly on DataFrames or Series. These approaches are faster and should be implemented unless specific operations need to be performed on individual rows or column elements.

The `.apply()` method allows one to apply a custom function element-wise to each row or column of a DataFrame or Series. The function can be defined separately or passed as an anonymous lambda function. This offers flexibility to define complex function logic for tasks that cannot be easily achieved with built-in functions. While `.apply()` is generally performance optimized, it might be slower than vectorized operations for simple calculations due to function call overhead.

Vectorized operations leverage built-in functions and methods that perform element-wise calculations on DataFrames or Series directly. Mathematical functions include arithmetic, comparison, and logical operators. Pandas methods include aggregation and statistics, membership testing, string manipulation, Boolean indexing, etc. They are generally faster than `.apply()` for simple calculations because they take advantage of optimized vectorized code compiled by Pandas.

An example satellite system is decomposed next with mass and quantity attributes for a complete parts list. A more detailed and comprehensive system decomposition could be contained in a large external file. Vectorized operations will be used to compute the total system mass considering mass and quantity for each component.

The vectorization computes the product of quantity and mass columns using element-wise multiplication `df['quantity'] * df['mass'])` creating a new Series for total mass of each component type. Then the total satellite mass is found with the Pandas `sum` function applied to the new Series.

```python
import pandas as pd

# satellite components data with mass and quantity
satellite_data = {
  "component": [
    "Structure", "Battery", "Solar Panel", "Camera", "Computer",
    "Thruster", "Communication Module", "Thermal Control System",
```

```
      "Deployment System", "Cables & Connectors"
    ],
    "mass": [
      530, 2045, 87, 528, 239,
      55, 370, 153, 277, 112
    ],
    "quantity": [
      1, 1, 4, 1, 1,
      4, 1, 1, 1, 1
    ]
}
df = pd.DataFrame(satellite_data)

# Compute the component mass using vectorized operations
df['component_mass'] = df['quantity'] * df['mass']

# Calculate the overall total mass
total_satellite_mass = df['component_mass'].sum()

# Display the DataFrame
print(df)

# Display the total satellite mass
print(f"\nTotal Satellite Mass: {total_satellite_mass}")
```

	component	mass	quantity	total_mass
0	Structure	530	1	530
1	Battery	2045	1	2045
2	Solar Panel	87	4	348
3	Camera	528	1	528
4	Computer	239	1	239
5	Thruster	55	4	220
6	Communication Module	370	1	370
7	Thermal Control System	153	1	153
8	Deployment System	277	1	277
9	Cables & Connectors	112	1	112

```
Total Satellite Mass: 4822
```

Vectorized operations are much faster, more memory-efficient, and more scalable versus iterating through rows manually. They handle entire columns or DataFrames at once with compiled C code, making them the preferred approach for most data manipulations in Pandas and particularly important for large datasets.

2.4 SCIPY

SciPy is a general purpose library for scientific computing with a wide range of tools for numerical optimization, statistics, interpolation, signal processing, linear algebra, integration, differential equations, and more [18]. The SciPy functions are segmented across modules, and are generally imported one module at a time and namespaced accordingly (as opposed to all Pandas functions imported at once as `pd`). Hence Table 2.5 showing common operations performed with SciPy includes the different module import names for usage.

There is some function overlap with NumPy and an identically named module `linalg` for linear algebra, but SciPy has more comprehensive functions. For example it has a wider range of probability distributions and advanced statistical functions. The `dir(module_name)` command can always be used to list the contained functions.

In the Table 2.5 examples the uppercase symbols X and Y denote lists or one-dimensional NumPy arrays, and `f` stands for an arbitrary defined function. An A stands for a two-dimensional NumPy array. See https://docs.scipy.org/doc/scipy/reference/ for a complete reference of the SciPy API. A small sample of simple use cases follows.

Table 2.5: Common SciPy Tasks

Task	Examples
Optimization	
Non-linear least squares	`from scipy.optimize import curve_fit`
	`popt, pcov = curve_fit(func, xdata, ydata)`
Linear least squares (deprecated)	`from scipy.linalg import lstsq`
	`x, residuals, rank, singular_values = ↵`
	`lstsq(A, b)`
Minimization algorithms	`from scipy.optimize import minimize`
	`res = minimize(objective_function, x0, ↵`
	`method='Nelder-Mead')`
Integration	
Numerical integration	`from scipy.integrate import quad`
	`integral, error = quad(func, a, b)`
Differential equations	`from scipy.integrate import solve_ivp`
	`solution = solve_ivp(f, time_span, y0)`
	`time = solution.time`
	`y = solution.y`
Interpolation	
Linear interpolation	`from scipy.interpolate import interp1d`
	`f = interp1d(X, Y)`
	`new_y = f(new_x)`

Continued on next page

Table 2.5: Common SciPy Tasks – Continued

Task	Examples

Spline interpolation	`from scipy.interpolate import ↙`
	 `UnivariateSpline`
	 `spline = UnivariateSpline(X, Y)`

Linear Algebra	
Matrix multiplication	`from scipy import linalg` `C = linalg.matmul(A, B)`
Matrix inverse	`A = linalg.inv(B)`
Solve linear equations	`X = scipy.linalg.solve(A, b)`
Eigenvalues and eigen- vectors	`eigenvalues, eigenvectors = linalg.eig(A)`
Statistics	
Descriptive statistics	`from scipy.stats import describe`
Probability distributions	`rv = stats.norm(mu, sigma)`
Hypothesis testing	`statistic, pvalue = stats.ttest_ind(a, b)`
Signal Processing	
Fast Fourier Transform	`from scipy.fft import fft, ifft`
Filtering	`from scipy.signal import filtfilt`
Convolution	`from scipy.signal import convolve`
Sparse Matrices	
Creating sparse matrices	`from scipy.sparse import csr_matrix`
Sparse matrix operations	Use similar syntax as dense matrices
Other Useful Functions	
Distance metrics	`from scipy.spatial.distance import cdist`
Interpolation in n dimen- sions	`from scipy.interpolate import ↙` `RegularGridInterpolator`
Image processing mod- ules	`from scipy import ndimage, skimage`

2.4.1   STATISTICS

A 95% confidence interval of manufacturing yield percentage is computed from random samples of production runs. The script first computes the mean and standard deviations with NumPy (the SciPy mean and standard are deprecated so NumPy is used instead). The SciPy `norm.interval` function is called where `loc` specifies the mean and `scale` specifies the standard deviation. The interval is returned as a tuple.

```
import numpy as np
import scipy.stats as stats

# Manufacturing yield samples (percent)
yield_data = [80.6, 99.0, 85.3, 83.8, 69.9, 89.6, 91.1 , 66.2, 91.2, ↙
    82.7, 73.5, 82.0, 54.0, 82.9, 75.9, 98.3, 107.2, 85.5, 79.1 , 84↙
    .3, 89.3, 86.3, 79.0, 92.3, 87.0]

print('Mean = ', np.mean(yield_data), 'Std. Dev. = ', ↙
    np.std(yield_data))

# Create 95% confidence interval for mean value
interval = stats.norm.interval(confidence=0.95, ↙
    loc=np.mean(yield_data), scale=np.std(yield_data))
print("95% confidence interval for mean = ", interval)
```

```
Mean =  83.84 Std. Dev. =  10.723842594891
95% confidence interval for mean =  (62.82165473813, 104.85834526186)
```

2.4.2   OPTIMIZATION

It is desired to minimize the cost of building holding tanks, where the material cost depends on the exterior surface area. The SciPy `optimize` module is used in Listing 2.3 to design a cylindrical tank for a specific volume. First, the `minimize` method is chosen, which is suitable for finding the minimum value of a scalar function with or without constraints. The optimization is then formulated by defining the objective function as the surface area dependent on radius and height, and the constraint for the total volume.

An initial guess for the radius and height of the tank is provided to start the optimization algorithm. The constraint is then defined as a dictionary with the type `eq` indicating an equality constraint, and specifying the function that ensures the volume of the tank is exactly 1000 units. Bounds are set for the radius and height to ensure both remain positive throughout the optimization. The optimization uses the Sequential Least Squares Quadratic Programming (SLSQP) method, which is well-suited for many constrained optimization problems.

The `result` returned when calling `minimize` is a special object that contains information about the outcome of the optimization process. The attribute for the `success` Boolean flag is checked, and then the optimized values are printed out. The `message` attribute is a string that provides a description of the exit status of the optimizer.

Listing 2.3   Tank Surface Area Optimization with SciPy

```
import numpy as np
from scipy.optimize import minimize

# Define the objective function (Surface Area)
```

```python
def surface_area(x):
    r, h = x # x contains [r, h]
    return 2 * np.pi * r**2 + 2 * np.pi * r * h # Total surface area ↙
        of cylinder

# Define the constraint function (Volume)
def volume_constraint(x):
    r, h = x # x contains [r, h]
    return np.pi * r**2 * h - 1000 # Volume must be exactly 1000

# Initial guess for the variables [r, h]
x0 = [1, 1]

# Define the constraint as a dictionary
constraint = {'type': 'eq', 'fun': volume_constraint}

# Define bounds for the variables (r > 0 and h > 0)
bounds = [(0.1 , None), (0.1 , None)] # Radius and height must be ↙
    positive

# Perform the optimization using 'SLSQP' method
result = minimize(surface_area, x0, method='SLSQP', bounds=bounds, ↙
    constraints=constraint)

# Output the results
if result.success:
    optimized_r, optimized_h = result.x
    print(f"Optimized radius: {optimized_r:.2f}")
    print(f"Optimized height: {optimized_h:.2f}")
    print(f"Minimum surface area: {result.fun:.2f}")
else:
    print("Optimization failed:", result.message)
```

```
Optimized radius: 5.42
Optimized height: 10.84
Minimum surface area: 553.58
```

The light transmittance characteristics of a new material need to be quantified. The SciPy `curve_fit` optimization method is used to fit an exponential decay curve to experimental data for the material in Listing 2.4. Light measurements are taken at fixed distance intervals and recorded into a CSV file with the structure:

```
distance,intensity
0.0,241.571
0.08,202.015
0.16,179.053
0.24,158.883
...
```

The `curve_fit` method takes a defined model function, x and y data points, then finds the optimal values for the function parameters (a, b, and c) that minimizes the difference between the function's output and actual data points. The fitted parameter values are stored in `popt` as a tuple.

The script first imports the experimental data with Pandas, uses Matplotlib to draw a scatter plot of the data alongside a continuous fitted curve, and displays the calibrated decay curve equation written with LaTeX.

Listing 2.4   Light Decay Curve Fitting Optimization with SciPy

```python
from scipy.optimize import curve_fit
import pandas as pd
import matplotlib.pyplot as plt

# Read CSV file and extract data.
data = pd.read_csv("light_measurements.csv")
distance = data["distance"]
intensity = data["intensity"]

# Define exponential light decay function
def func(x, a, b, c):
  return a * np.exp(-b * x) + c

# Get curve fit optimal parameters (popt) and covariance matrix (pcov).
popt, pcov = curve_fit(func, distance, intensity)

# Extract fitted parameter values.
a, b, c = popt

# Create figure and axes.
fig, ax = plt.subplots()

# Plot data and fitted curve
ax.scatter(distance, intensity, label='Experimental Data')
ax.plot(distance, func(distance, *popt), 'r-', label='Fitted Curve')

# Create string to display LaTeX equation with parameters.
param_text = f"$f(x) = {a:.2f} e^{{-{b:.2f}x}} + {c:.2f}$"

# Add text box for equation.
ax.text(1.5, 100, param_text, ha='left', va='top', fontsize=10, ⤶
    bbox=dict(facecolor='none', edgecolor='none', alpha=0.5))

ax.set(xlabel='Material Distance (cm)', ylabel='Light Intensity ⤶
    (lumens)')
plt.legend(loc='upper center')
```

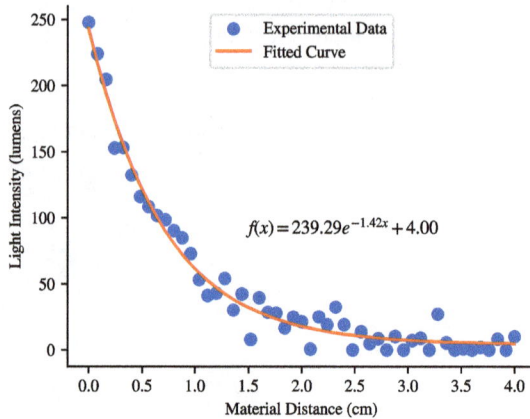

Figure 2.13   Exponential Decay Curve Fit

2.4.3   HYPOTHESIS TESTING

A hypothesis test of the exponential model fit can be performed with the SciPy `ttest_ind` module. It calculates the T-test for the means of two independent samples of scores, assuming the null hypothesis that the samples have identical average values. Here the residuals are compared to a normal distribution with a mean of zero.

```python
# t-test for independent samples
from scipy.stats import ttest_ind

# Significance level
alpha = 0.05

# Calculate residuals as difference between observed and fitted values
residuals = intensity - func(distance, *popt)

# Perform t-test against zero residuals assuming normally ↙
    distributed errors
t_statistic, p_value = ttest_ind(residuals, np.zeros_like(residuals))

print(f"t-statistic: {t_statistic:.4f}")
p_value = abs(p_value) # two-tailed test
print(f"p-value (two-tailed): {p_value:.4f}")

if p_value > alpha:
    print("Fail to reject the null hypothesis that the model fits ↙
        the data (residuals are zero).")
else:
    print("Reject the null hypothesis that the model fits the data.")
```

```
t-statistic: -0.0000
p-value (two-tailed): 1.0000
Fail to reject the null hypothesis that the model fits the data
(residuals are zero).
```

2.4.4 SIGNAL PROCESSING

Scipy provides functions for analysis and manipulation of signals, such as sound, images, and other data including `convolve`, `correlate`, `spectrogram`, and `lfilter`. This example uses `spectrogram` to visualize the frequency content of a signal. A known set of sine waves is created in this case in order to see how they are visualized in the spectrogram image.

The NumPy `linspace` creates 1000 evenly spaced timepoints, which also sets the size of the corresponding x signal array. The signal is a combination of six sine waves with chosen frequencies to create a visually interesting spectrogram. The `cmap='inferno'` argument in the pcolormesh function specifies the color map for the plot, and `10 * np.log10(Sxx)` is used to convert the spectrogram to a logarithmic scale to enhance the visualization. The `plt.colorbar` adds a color bar to the plot to indicate the power spectral density in decibels.

```python
import numpy as np
from scipy import signal
import matplotlib.pyplot as plt

# Create a signal with 6 sine waves
fs = 1000
t = np.linspace(0, 1, fs, endpoint=False)
x = (
    5 * np.sin(2*np.pi*10*t) +
    3 * np.sin(2*np.pi*50*t) +
    2 * np.sin(2*np.pi*100*t) +
    4 * np.sin(2*np.pi*200*t) +
    1 * np.sin(2*np.pi*300*t) +
    6 * np.sin(2*np.pi*500*t)
)

# Compute the spectrogram
f, t, Sxx = signal.spectrogram(x, fs)

# Plot the spectrogram with colorbar
fig, ax = plt.subplots()
spectrogram_image = ax.pcolormesh(t, f, 10 * np.log10(Sxx), ⤶
    shading='auto', cmap='inferno')
fig.colorbar(spectrogram_image, label='Power Spectral Density [dB]')
ax.set(ylabel='Frequency (Hz)', xlabel='Time (sec)')
```

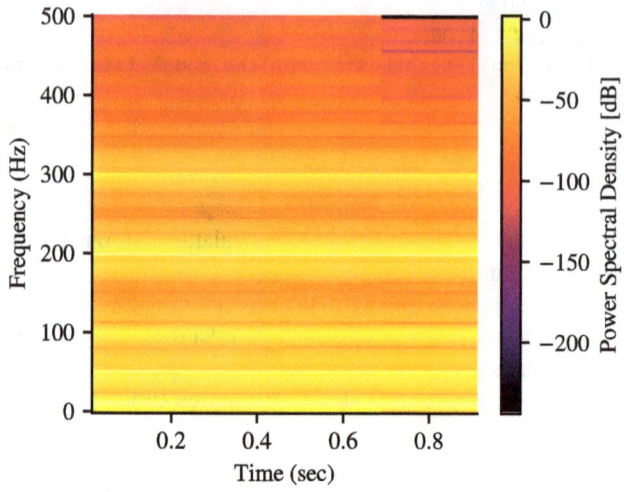

Figure 2.14   Signal Spectrogram

2.5   NETWORKX

The NetworkX library is widely used to create, modify, and analyze network models for engineering systems or processes involving multiple components, connections, or interactions. Network data structures are particularly useful for modeling, analyzing, and optimizing such systems because they represent them as graphs consisting of nodes (vertices) and edges (connections) between nodes. In a graph, nodes represent system entities, and edges represent relationships between those entities.

NetworkX includes algorithms to analyze graph, node, and edge properties such as connectivity, flow, and centrality. Paths through networks can be computed and optimized. These analyses can help identify bottlenecks, optimize routes, and improve overall system performance in areas such as communication networks, electrical circuits, supply chains, and project management.

Constructing a network in NetworkX fundamentally involves adding nodes and edges. There is flexibility to add them individually, in groups, or from external data sources such as lists or datasets. Additionally, attributes can be assigned to graphs, nodes, and edges to provide contextual or quantitative data for computations. NetworkX uses a default *weight* attribute for graph computations, though any named attribute can be used, as is demonstrated later.

Table 2.6 provides examples of common NetworkX operations. It includes functions for creating different types of network graphs, adding nodes and edges, computing various network properties and algorithms such as shortest paths, maximum flow, and centrality measures, as well as visualization and file I/O. Full documentation is available at https://networkx.org/documentation/stable/reference/index.html, with additional resources on graph theory and network analysis at https://networkx.org/nx-guides/index.html.

2.5.1   GRAPH TYPES

Graphs can be undirected, or directed, or allow multiple edges between the same pair of nodes, depending on the system being modeled. NetworkX supports the following types of graphs:

- **Undirected graph** using `nx.Graph()` is a regular graph where edges between nodes are undirected. An edge connection can be traversed in both directions, and each pair of nodes can only have one edge. This type of graph is suitable for problems where the relationship between nodes is bidirectional, such as a road network or electrical grid with flows in both directions.

- **Directed graph** using `nx.DiGraph()` contains edges with specific directions. An edge between nodes can only be traversed in one direction, but a separate edge can be explicitly added for traversal in the other direction. This graph type models one-way relationships, such as communication networks where data flows in one direction, electrical circuits with directed currents, or sequential task dependencies. They are visualized with directional arrows.

Table 2.6

Common NetworkX Operations

Task	Examples
Graph Creation	
Create a regular graph	`G = nx.Graph()`
Create a directed graph	`G = nx.DiGraph()`
Create a multigraph	`G = nx.MultiGraph()`
Add a single node	`G.add_node('A')`
Add multiple nodes	`G.add_nodes_from([2, 3])`
Add nodes with attributes	`G.add_node('A', role='Router')`
Add a single edge	`G.add_edge(1, 2)`
Add multiple edges	`G.add_edges_from([('A', 'B'), ('B', 'C')])`
Add edges with attributes	`G.add_edge('A', 'B', latency=10, ⤶ label='Wi-Fi')`
	`G.add_edge(3, 4, {'weight': 20, ⤶ 'label': 'Route1'})`
Import nodes/edges from a list	`G.add_edges_from(edge_list)`
	`G.add_nodes_from(node_list)`
Add node or edge attributes	`G.add_node(1, label="Start"),`
	`G.add_edge(1, 2, weight=10)`
Get nodes and edges	`G.nodes(), G.edges()`
Get the degree of a node	`G.degree(1)`
Graph Operations	
Find all paths between nodes	`list(nx.all_simple_paths(G, 1, 3))`
Find the shortest path	`nx.shortest_path(G, 1, 3)`
Calculate total path weight	`nx.path_weight(G, path, ⤶ weight='attribute')`
Check if nodes are connected	`nx.has_path(G, 1, 3)`
Find all neighbors of a node	`list(G.neighbors(1))`
Graph Analysis	
Clustering coefficient	`nx.clustering(G, 1)`
Degree centrality	`nx.degree_centrality(G)`
Betweenness centrality	`nx.betweenness_centrality(G)`
Find connected components	`nx.connected_components(G)`
Layout and Visualization	
Compute layout positions	`pos = nx.spring_layout(G)`
Draw the graph	`nx.draw(G, pos)`
Draw with labels	`nx.draw(G, pos, with_labels=True)`
File Input and Output	
Read graph from a file	`G = nx.read_edgelist('edges.txt')`
Write graph to a file	`nx.write_edgelist(G, 'output.txt')`
Graph Properties	
Check graph density	`nx.density(G)`
Get graph diameter	`nx.diameter(G)`
Check if graph is connected	`nx.is_connected(G)`

- **Undirected multigraph** using nx.MultiGraph() allows multiple edges between the same pair of nodes. An edge connecting two nodes can represent different types or layers of connections. This type is useful in cases such as transportation networks, where multiple paths connect two physical locations, or parallel systems with different connections between the same entities. When visualized, multiple edges between nodes overlap as a single line.

- **Directed multigraph** using nx.MultiDiGraph() allows multiple directed edges between the same pair of nodes. This type models systems with multiple directed relationships, such as distinct communication channels between devices with varying capacities, bandwidths, latencies, or costs. These are visualized with directional arrows, but multiple edges overlap as a single line.

2.5.2 GRAPH CREATION AND VISUALIZATION

NetworkX provides flexible methods for adding nodes and edges to a graph. Per Table 2.6, nodes can be added one at a time or as a group from any iterable such as a list. Similarly, edges can be added individually, or in groups. Attributes can be assigned to nodes or edges using keyword arguments or dictionaries. Nodes and edges can also be imported from external lists or files.

The example in Listing 2.5 models a Local Area Network (LAN) where nodes represent devices (modems, routers, and switches), and edges represent bi-directional communication links between them. In this example, a graph is initialized using nx.Graph(), nodes are added for the devices, and edges are defined for the communication links.

After creating the graph, it can be visualized as shown in Figure 2.15 using the nx.draw() function as an undirected network. The graph rendering illustrates the structure of the nodes and edges, aiding in the understanding of the system layout. The graph is rendered internally using Matplotlib.

The default positioning algorithm produces random layout patterns, which can vary across runs. More control over the layout can be achieved using the optional pos argument, as discussed later. However, for better control of hierarchical structures and directed graphs, Graphviz is recommended, as explained in Section 2.6. It also offers more advanced styling and customization options.

Listing 2.5 NetworkX Undirected Graph

```python
import networkx as nx
import matplotlib.pyplot as plt

# Create undirected graph
G = nx.Graph()

# Add nodes for devices
G.add_nodes_from(['Modem', 'Router1', 'Router2', 'Switch1', ↙
    'Switch2', 'Switch3'])
```

```
# Add edges for communication links between devices
G.add_edges_from([('Modem', 'Router1'),
                  ('Router1', 'Router2'),
                  ('Router1', 'Switch1'),
                  ('Router2', 'Switch2'),
                  ('Router2', 'Switch3')])

# Visualize the graph
nx.draw(G, with_labels=True)
```

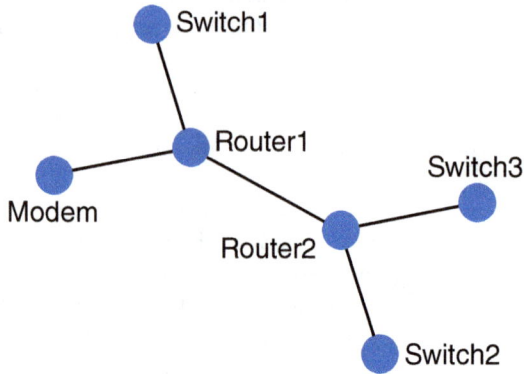

Figure 2.15   LAN Communication Network

The next example in Listing 2.6 demonstrates a network model for optimizing a circuit board manufacturing process as a directed graph. This is applied to the fictional FastCircuit company that designs, fabricates, manufactures, and refurbishes circuit board electronics. The network graph models the physical flow of circuit boards from an initial stock of substrate boards through successive workstations where various operations are performed. The nodes represent physical workstation locations, while the edges represent transit paths between them, characterized by average duration times due to physical layout and transport methods. The duration attributes are added for analysis in Section 2.5.4.

The default layout in Figure 2.16 shows a random arrangement of the nodes. However, a more desirable layout would be a directed graph from left to right, providing a clearer visualization of the process sequence. This can be achieved using Graphviz, as discussed in Section 2.6.

Listing 2.6   NetworkX Directed Graph

```
import networkx as nx

# Create a directed graph
G = nx.DiGraph()
```

```
# Add edges and nodes
G.add_edge("Board Stock", "Fabrication")
G.add_edge("Fabrication", "Routing")
G.add_edge("Routing", "Assembly 1")
G.add_edge("Routing", "Assembly 2")
G.add_edge("Assembly 1", "Test Station 1")
G.add_edge("Assembly 1", "Test Station 2")
G.add_edge("Assembly 2", "Test Station 1")
G.add_edge("Assembly 2", "Test Station 2")
G.add_edge('Test Station 1', 'Packaging')
G.add_edge('Test Station 2', 'Packaging')

# Visualize the graph
nx.draw(G, with_labels=True)
```

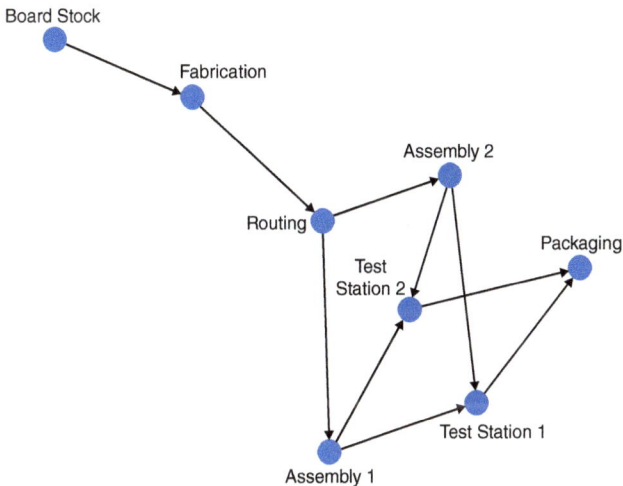

Figure 2.16   Circuit Board Manufacturing Path Network

The next reference example in Listing 2.7 models the structure of a satellite system design as a directed hierarchy. This tree hierarchy represents and communicates the system's engineering design, project work allocation, and computation of weight and cost based on its structural decomposition at different levels. It is directed down from the top level illustrating the decomposition and serving as a natural structure for roll-up calculations. The script adds edge pairs and visualizes the graph in Figure 2.17.

The displayed graph using the default layout places the top node of the hierarchy in the center. Radial connections spread out as successive levels in the hierarchy. It can be visualized which main branches go down one or two levels of decomposition by counting the connections. However, a directed tree visualization is desirable.

Note that there is an imperfection due to a long label being clipped off. NetworkX adjusts layouts for node placement only. This case can be remedied with Matplotlib parameters to add margin (used in Section 2.5.3).

Listing 2.7   NetworkX Directed Hierarchical Graph

```python
import networkx as nx

# Create directed graph
G = gv.Digraph()

# add nodes and edges
G.add_edges_from([('Satellite', 'Payload'),
        ('Satellite', 'Communication'),
        ('Satellite', 'Power'),
        ('Satellite', 'Propulsion'),
        ('Satellite', 'Attitude Control'),
        ('Satellite', 'Computer'),
        ('Payload', 'Camera'),
        ('Communication', 'Antenna'),
        ('Communication', 'Transceiver'),
        ('Power', 'Solar Panels'),
        ('Power', 'Battery'),
        ('Propulsion', 'Thrusters'),
        ('Attitude Control', 'Reaction Wheels'),
        ('Attitude Control', 'Magnetorquers')])

# Visualize the graph
nx.draw(G, with_labels=True)
```

2.5.3   GRAPH LAYOUTS

There are several layout options for positioning nodes and edges within a graph. The layout determines the coordinates of nodes in a two-dimensional plane. Common layout functions in NetworkX include the following. A complete list of graph layouts is at https://networkx.org/documentation/stable/reference/drawing.html#module-networkx.drawing.layout.

- spring_layout is a layout where node positions are computed using a force-directed algorithm where nodes repel each other, while edges act as springs to keep connected nodes close. This often results in an aesthetically pleasing layout, but the node locations are random unless a seed is provided. This is the default.

- circular_layout places the nodes in a circle, useful for highlighting cyclic structures.

- shell_layout arranges nodes in concentric circles, useful for layered graph structures.

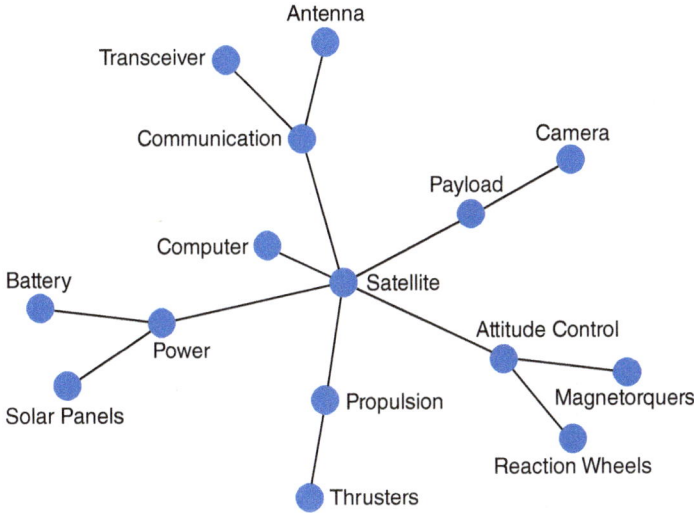

Figure 2.17   Satellite Design

- `spectral_layout` is based on the eigenvectors of the graph Laplacian matrix; it positions nodes in a way that minimizes a quadratic energy function.

- `random_layout` places nodes randomly within a unit square.

In order to compare the common layout algorithms for different system types, the function in Listing 2.8 was used to generate different layouts for the reference models in Table 2.7. The LAN network model is an undirected network, the manufacturing process flow model represents sequential activities, and the satellite design is a hierarchical structure of system components.

Per Table 2.7, the default spring layouts evenly spread the nodes with random directions. The hierarchical system is seen with the top node in the middle with radial edges. The circular layouts put all nodes in a ring, and in these cases the shell layout produces mirror images of the circular (which isn't the case for all networks). The spectral layout minimizes an energy function but places some nodes on top of each other. The occluding is undesirable for visualizing these situations. The beginning and end nodes of directed graphs are difficult to follow in some of the cases.

Listing 2.8   Plotting NetworkX Layouts

```
import matplotlib.pyplot as plt

def plot_layouts(G, title):
    for layout in ['spring_layout', 'circular_layout', ∠
        'shell_layout', 'spectral_layout']:
        fig, ax = plt.subplots()
```

```
    nx.draw(G, pos=getattr(nx, layout)(G), ↵
        node_color='lightblue', with_labels=True, width=2)
    plt.savefig(f"{title} {layout}.pdf")

plot_layouts(G, "LAN network")
```

Table 2.7

NetworkX Layout Examples

LAN Network	**Manufacturing Flow**	**Satellite Design**

spring

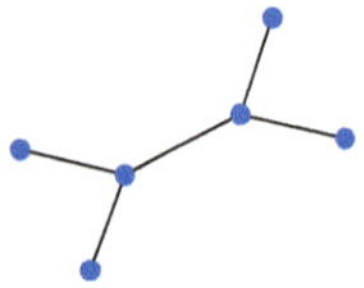

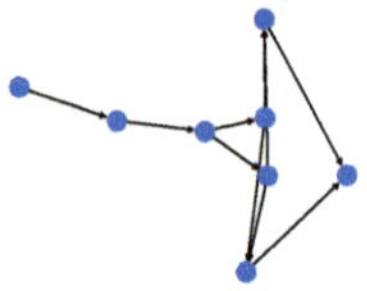

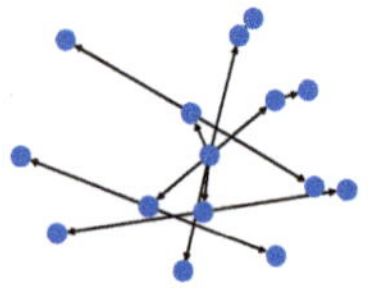

circular

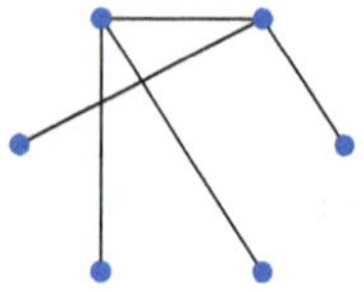

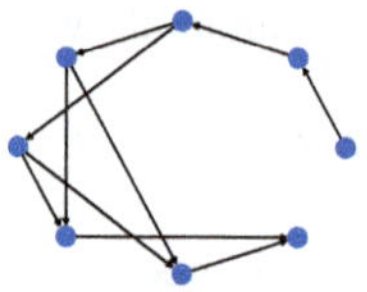

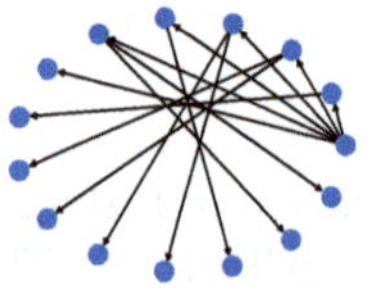

shell

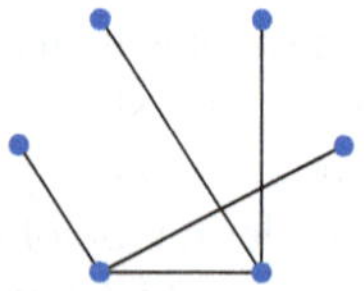

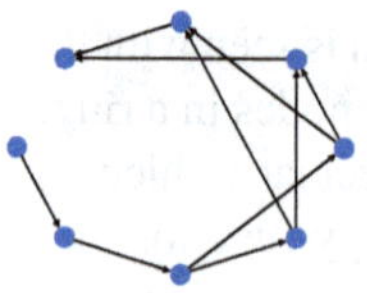

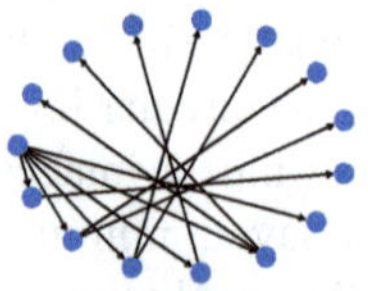

spectral

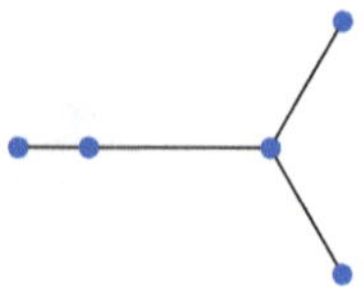

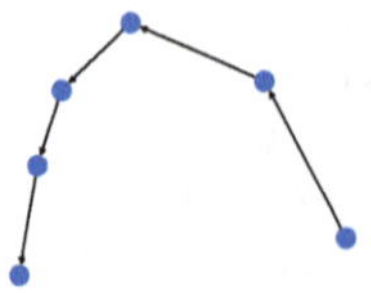

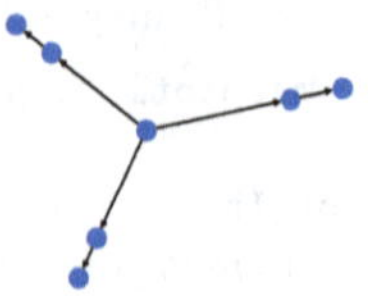

The node positions in layouts such as `spring_layout` are random due to the initial conditions used in force-directed algorithms. These algorithms are iterative, calculating forces between nodes and edges to determine the final layout. As a result, the outcome can vary between runs unless a seed is explicitly provided. Layouts like `random_layout` are designed to place nodes randomly by default.

The optional `pos` argument in the `nx.draw` function allows for specifying exact node positions, offering more control over the layout. By default, NetworkX automatically generates node positions based on the selected layout algorithm (e.g., `spring_layout`). For simple visualizations, the `pos` argument can often be omitted. The following example demonstrates how to generate a consistent layout by providing a fixed seed to the `nx.spring_layout()` function for the randomization:

```python
# Visualize the undirected graph
pos = nx.spring_layout(G, seed=333)
nx.draw(G, pos, with_labels=True)
```

NetworkX provides basic functionality for visualizing all types of graphs, but its primary purpose is graph analysis rather than visualization. In the future, graph visualization features may be removed from NetworkX or made available as an add-on package. For more advanced graph visualization, using a dedicated tools such as Graphviz described in Section 2.6 is recommended.

2.5.4   GRAPH ANALYSIS

Graphs can be analyzed in many ways. NetworkX provides a wide range of graph algorithms and calculations that are useful in engineering applications, such as optimizing routes, identifying hubs or bottlenecks, and assessing the importance of nodes or edges. Some useful algorithms include:

- **Shortest path** using `shortest_path()` finds the minimum distance or number of steps required to traverse from one node to another as the most efficient route.

- **Maximum flow** with `maximum_flow()` calculates the maximum possible flow between two nodes in a network, which is useful in optimizing network throughput.

- **Degree centrality** using `degree_centrality()` measures node importance based on connections, identifying hubs in transportation networks

- **Betweenness centrality** using `betweenness_centrality()` finds nodes that control information flow, identifying bottlenecks

- **Clustering coefficient** with `clustering_coefficient()` measures the degree to which nodes cluster together, which can be useful in analyzing the robustness of networks.

- **Connectivity** using `is_connected()` checks whether nodes in a network are connected.

- **Connected components** using `connected_components()` identifies isolated subgraphs or clusters within a larger network.

One common task in engineering is finding the shortest path between two nodes in a network. In a communication network, for instance, the shortest path represents the minimum number of links (edges) required to transmit data between two devices (nodes). The following example for the LAN network model finds the shortest path between `Router1` and `Switch2` using the `nx.shortest_path` function. It returns a list of nodes along the path with the minimum number of hops. In this simple case, there is a direct route between the nodes, with one alternative path involving three hops.

```python
# Find the shortest path between Router1 and Switch2
shortest_path = nx.shortest_path(G, 'Router1', 'Switch2')
print(f"The shortest path from Router1 to Switch2 is {shortest_path}")
```
```
The shortest path from Router1 to Switch2 is ['Router1', 'Switch2']
```

The example in Listing 2.9 calculates basic measurements for each node in the LAN model and outputs a summary table.

Listing 2.9   NetworkX Node Measures

```python
import networkx as nx
import pandas as pd

def calculate_node_measures(G):
  """Calculates degree, betweenness, and clustering coefficients for ↙
      each node in a graph and returns a combined pandas DataFrame.

  Args:
    G: A NetworkX graph.

  Returns:
    A pandas DataFrame containing node names, degree centrality, ↙
        betweenness centrality, and clustering coefficient values.
  """

  degree_centrality = nx.degree_centrality(G)
  betweenness_centrality = nx.betweenness_centrality(G)
  clustering_coefficient = nx.clustering(G)

  # Combine the results into a single DataFrame
  data = {'Node': list(G.nodes()),
          'Degree Centrality': [degree_centrality[node] for node in ↙
              G.nodes()],
          'Betweenness Centrality': [betweenness_centrality[node] for ↙
              node in G.nodes()],
          'Clustering Coefficient': [clustering_coefficient[node] for ↙
              node in G.nodes()]}
```

```
df = pd.DataFrame(data)
df.sort_values(by='Degree Centrality', ascending=False, inplace=True)

return df

# LAN model graph
G = nx.Graph()
G.add_edges_from([('Modem', 'Router1'),
                  ('Router1', 'Router2'),
                  ('Router1', 'Switch1'),
                  ('Router2', 'Switch2'),
                  ('Router2', 'Switch3')])

# Calculate node measures
node_measures = calculate_node_measures(G)
print(node_measures)
     Node  Degree Centrality  Betweenness Centrality  Clustering Coefficient
1  Router1                0.6                     0.7                       0
2  Router2                0.6                     0.7                       0
0    Modem                0.2                     0.0                       0
3  Switch1                0.2                     0.0                       0
4  Switch2                0.2                     0.0                       0
5  Switch3                0.2                     0.0                       0
```

2.5.5   PATH COMPUTATIONS AND EDGE ATTRIBUTES

Computing the weight or cost of a path is crucial in many engineering applications. This can be achieved using edge attributes. Edges in a graph may have attributes such as weight, duration, bandwidth, latency, or transmission cost. By assigning a weight or another relevant attribute to each edge, the total cost of traversing a specific path can be calculated.

In Listing 2.10, the communication network is enhanced by assigning latency (in milliseconds) to each transmission link. The nx.path_weight() function is used to calculate the total latency of a path based on the specified edge attribute. Each edge is added separately with G.add_edge to accommodate the attribute values. Nodes are inherently defined within the G.add_edge function parameters and don't need to be added separately.

Listing 2.10   NetworkX Graph Path Computations

```
import networkx as nx

# Create undirected graph
G = nx.Graph()

# Add edges with latencies (costs)
G.add_edge('Router1', 'Router2', latency=10)
G.add_edge('Router2', 'Switch1', latency=15)
G.add_edge('Switch1', 'Switch2', latency=20)
```

```python
G.add_edge('Router1', 'Switch2', latency=30)

start = 'Router1'
end = 'Switch1'

# Find all paths between start and end nodes
all_paths = list(nx.all_simple_paths(G, start, end))
print(f"All paths from {start} to {end}: \n{all_paths}")

# Find the shortest path and compute its total cost (latency)
path = nx.shortest_path(G, start, end)
path_cost = nx.path_weight(G, path, weight='latency')
print(f"Shortest path from {start} to {end} is {path}")
print(f"Total latency for this path = {path_cost} ms")
```
```
All paths from Router1 to Switch1:
[['Router1', 'Router2', 'Switch1'], ['Router1', 'Switch2', 'Switch1']]
Shortest path from Router1 to Switch1 is ['Router1', 'Router2', 'Switch1']
Total latency for this path = 25 ms
```

The next example in Listing 2.11 seeks to optimize the FastCircuit circuit board manufacturing process through path analysis. A custom edge attribute for duration is added. With this, the total transit time between nodes can be calculated. All possible paths from the start to the end are identified using `nx.all_simple_paths()`, and their total durations are calculated using `nx.path_weight()`, specifying the attribute name `duration`.

The default layout in Figure 2.16 shows a random arrangement of the nodes, which may be sufficient if only path durations are required. However, a more desirable layout would be a directed graph from left to right, providing a clearer visualization of the process sequence. This can be achieved using Graphviz, as discussed in Section 2.6.

Listing 2.11 NetworkX Directed Graph Path Computations

```python
import networkx as nx

# Create a graph object
G = nx.DiGraph()

# Manufacturing process
# Add edges with duration attribute
G.add_edge("Board Stock", "Fabrication", duration=4)
G.add_edge("Fabrication", "Assembly 1", duration=3)
G.add_edge("Fabrication", "Assembly 2", duration=3)
G.add_edge("Fabrication", "Assembly 3", duration=5)
G.add_edge("Assembly 1", "Test Station 1", duration=3)
G.add_edge("Assembly 1", "Test Station 2", duration=5)
G.add_edge("Assembly 2", "Test Station 1", duration=3)
G.add_edge("Assembly 2", "Test Station 2", duration=6)
G.add_edge("Assembly 3", "Test Station 1", duration=4)
```

```
G.add_edge("Assembly 3", "Test Station 2", duration=5)
G.add_edge('Test Station 1', 'Packaging', duration=2)
G.add_edge('Test Station 2', 'Packaging', duration=2)

# Find paths and their durations
paths = nx.all_simple_paths(G, 'Board Stock', 'Packaging')
print('Path Durations')
for path in paths:
    path_duration = nx.path_weight(G, path, 'duration')
    print(path, path_duration)

# Find shortest path
shortest_path = nx.shortest_path(G, 'Board Stock', 'Packaging', ↙
    weight='duration')
print("Shortest path = ", shortest_path)
print("Duration = ", nx.path_weight(G, shortest_path, 'duration'))

nx.draw(G, with_labels=True)
Path Durations
['Board Stock', 'Fabrication', 'Assembly 1', 'Test Station 1', 'Packaging'] 12
['Board Stock', 'Fabrication', 'Assembly 1', 'Test Station 2', 'Packaging'] 14
['Board Stock', 'Fabrication', 'Assembly 2', 'Test Station 1', 'Packaging'] 12
['Board Stock', 'Fabrication', 'Assembly 2', 'Test Station 2', 'Packaging'] 15
['Board Stock', 'Fabrication', 'Assembly 3', 'Test Station 1', 'Packaging'] 15
['Board Stock', 'Fabrication', 'Assembly 3', 'Test Station 2', 'Packaging'] 16
Shortest path =  ['Board Stock', 'Fabrication', 'Assembly 1', 'Test Station 1',
'Packaging']
Duration =  12
```

Node and edge attributes can be useful for many calculations. NetworkX has ample facilities for computing with edge data, but it doesn't provide any functions for quantitative node attributes. Network node calculations are demonstrated in the project network critical path application in Section 3.7.

2.5.6 CUSTOMIZATION

Customization of graph visualizations allows for a clearer and more meaningful display of complex network structures. Visual properties such as node color, edge width, edge styles, and labeling are crucial to enhancing graph readability, especially in networks that involve multiple layers of information. Customizing node colors helps differentiate between various types of nodes or groupings. Edge widths and styles can reflect edge attributes such as weight or different connection types.

NetworkX requires handling various elements of the graph, such as node positions, colors, and edge styles, in multiple steps. Dedicated graph visualization tools like Graphviz have more straightforward built-in support, and NetworkX visualizations can be improved with Graphviz due to its more extensive capabilities.

The `nx.draw` function serves as the primary tool for drawing graphs. Customization involves passing specific parameters to `nx.draw` to control visual aspects of nodes and edges. The `node_color` parameter is used to assign colors to the nodes. It can be based on node attributes or other criteria, such as node names or types. The edge width and style parameters allow control over the appearance of edges.

Node labels are automatically handled by `nx.draw`, while edge labels require the use of `nx.draw_networkx_edge_labels`. The `pos` argument is necessary for placing both nodes and edge labels, as it specifies the layout of the graph. Without it, the labels will not display in the correct positions.

The overall positional layout of the graph is also managed by the `pos` parameter, which is generated by the layout algorithms. Layout is used to calculate the positions of the nodes and edges for proper rendering. For example, `pos = ↙ nx.spring_layout(G)` will calculate the node positions.

In the example Listing 2.18 for the LAN network, components are customized via node colors, edge styles, and edge labels. This makes a more informative graph that differentiates between routers and switches, as well as WiFi and Ethernet connections, while also displaying connection latencies.

Nodes are assigned colors based on their type, whether routers or otherwise, using a conditional list comprehension. The edges are styled based on the type of connection by using a conditional list comprehension that checks the label attribute of each edge. The `pos` parameter generated with `nx.spring_layout(G)` is required in `nx.draw_networkx_edge_labels` to specify where to place the edge labels.

Listing 2.12 NetworkX Customization

```python
import networkx as nx
import matplotlib.pyplot as plt

# Create an undirected multigraph
G = nx.Graph()
G.add_edge('Router1', 'Router2', label='WiFi', latency=6)
G.add_edge('Router1', 'Switch1', label='ethernet', latency=.3)
G.add_edge('Router2', 'Switch2', label='ethernet', latency=.3)
G.add_edge('Switch1', 'Switch2', label='ethernet', latency=.2)
G.add_edge('Switch2', 'Switch3', label='ethernet', latency=.2)

# Draw graph with customized node colors and edge styles
pos = nx.spring_layout(G) # Calculate node positions
node_colors = ['lightblue' if 'Router' in node else 'orange' for ↙
    node in G.nodes()]
edge_styles = ['dashed' if d['label'] == 'WiFi' else 'solid' for u, ↙
    v, d in G.edges(data=True)]
edge_colors = ['green' if d['label'] == 'WiFi' else 'blue' for u, v, ↙
    d in G.edges(data=True)] # Set edge colors
nx.draw(G, pos, with_labels=True, node_color=node_colors, ↙
    style=edge_styles, edge_color=edge_colors, width=2)

# Extract edge labels from 'label' attribute and draw them
edge_labels = {(u, v): f"{d['label']} ({d['latency']} ms)" for u, v, ↙
    d in G.edges(data=True)}
nx.draw_networkx_edge_labels(G, pos, edge_labels=edge_labels)
```

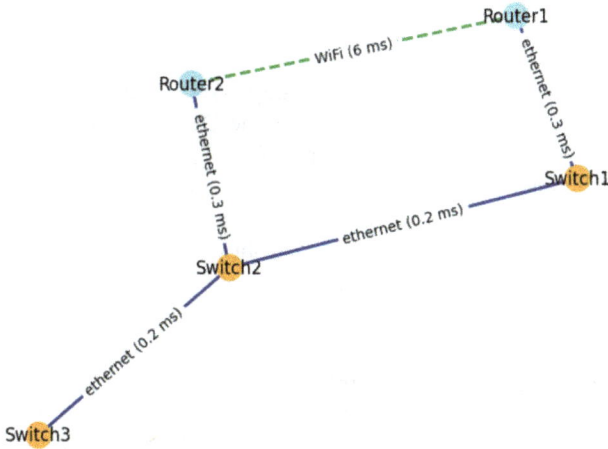

Figure 2.18   NetworkX Customization

While NetworkX offers flexible and customizable layout options, it has limitations when creating highly optimized and visually structured graphs compared to
dedicated graph visualization tools. Graphviz has more advanced algorithms to generate cleaner and more organized layouts, particularly for large or complex graphs.
In some cases, using Graphviz described in Section 2.6 becomes necessary due to
the limitations of NetworkX.

2.6 GRAPHVIZ

Graphviz is an extensive library for creating and visualizing graph diagrams. It can generate visual representations of various graph types, including directed and undirected graphs, trees, and flowcharts. It uses the DOT language for describing the graph, which is then rendered into an image file.

While both Graphviz and NetworkX are used for graph manipulation and visualization, Graphviz excels in rendering, offering more advanced visuals. In contrast, only NetworkX has graph computational capabilities. The common graph structure as a network of nodes and edges enables compatibility between them.

Graphviz provides a simple, intuitive syntax for creating graph diagrams. It supports a wide range of graph types, making it versatile for many applications, including system modeling, data visualization, flowchart creation, and network diagramming. The library allows for extensive customization of the graph's appearance, including colors, shapes, and labels, to meet specific visualization needs. Additionally, it integrates well with other libraries and tools.

Table 2.8 summarizes common Graphviz operations for creating, modifying, visualizing, and managing graph layouts. A complete API reference is available at https://graphviz.readthedocs.io/en/stable/api.html.

2.6.1 GRAPH TYPES

Graphviz supports both undirected and directed graphs, with multiple edges between the same pair of nodes allowed by default. Unlike NetworkX, no additional configuration is needed to handle multigraphs. The graphs can be either cyclic or acyclic. Graph types in Graphviz include:

- **Undirected graph** using `gv.Graph()` defines a regular graph where edges between nodes are undirected. An **undirected multigraph** can be created by specifying multiple edges between node pairs.

- **Directed graph** using `gv.Digraph()` creates a graph where edges are directional. A **directed multigraph** can be created with multiple directed edges between node pairs.

The decision of which type of graph to use depends on the system being modeled, whether data or entities flow in a single direction or are undirected. Not all layout options are compatible with both types as described in Section 2.6.3.

2.6.2 GRAPH CREATION AND VISUALIZATION

In the next example in Listing 2.13, Graphviz is applied to visually illustrate the initial structure of a satellite system design as a hierarchy in Figure 2.19. This hierarchy represents and communicates the system's engineering design, project work allocation, and computation of weight and cost based on its structural decomposition.

The system decomposition is visualized as a directed graph, specified by the `gv.DiGraph()` method. It is directed because the hierarchy represents a tree structure

Table 2.8

Common Graphviz Operations

Task	Examples
Graph Creation	
Create a new graph	`import graphviz as gv`
Initialize directed graph	`dot = gv.Digraph()`
Initialize undirected graph	`dot = gv.Graph()`
Set graph attributes	`dot.attr(rankdir='LR')`
	`dot = gv.Graph(graph_attr={'rankdir': ⤸`
	`'LR'}, node_attr={'shape': 'none'}, ⤸`
	`edge_attr={'penwidth': '2'})`
	`dot.attr('node', fontname="arial", ⤸`
	`fontcolor='blue', color='invis')`
	`dot.edge_attr.update(color="gray50", ⤸`
	`arrowsize="0.5")`
Node and Edge Operations	
Add node	`dot.node('A')`
Add multiple nodes	`dot.node('B'), dot.node('C')`
Add edge between nodes	`dot.edge('A', 'B')`
Add multiple edges	`dot.edges(['AB', 'BC', 'CA'])`
Set node attributes	`dot.node('A', shape='circle')`
Set edge attributes	`dot.edge('A', 'B', label='cost')`
Layout and Visualization	
Generate layout	`dot.render('output', format='png')`
Display graph	`dot.view()`
Set layout engine	`dot.engine = 'circo'`
Graph Properties	
Set global node properties	`dot.attr('node', shape='box', color='red')`
Set global edge properties	`dot.attr('edge', style='dashed')`
Set rank for nodes	`dot.attr(rank='same')`
File Input and Output	
Save graph to file	`dot.save('graph.dot')`
Read graph from file	`Graph(filename='graph.dot').source`

with successive levels decomposed from a parent node. Nodes and edges can be
added individually, and the graph can then be visualized as shown below.

Listing 2.13   Graphviz Graph

```python
import graphviz as gv

# Create directed graph
G = gv.Digraph()

# Add nodes and edges
G.node('Satellite')
G.node('Payload')
G.node('Power')
G.edge('Satellite', 'Payload')
G.edge('Satellite', 'Power')

# Display graph
G
```

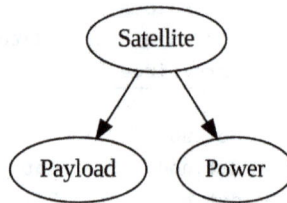

Figure 2.19   Initial Satellite Component Tree

In the next revision in Listing 2.14, more layers, nodes, and edges are added
to fully represent the satellite system structure. The G.edges() function for adding
edges implicitly adds the nodes in defined edge pairs, reducing redundancy in defining nodes separately using G.node(). An external list of edges could also be passed
to G.edges() to draw the entire tree. The resulting graph, shown in Figure 2.20,
illustrates the hierarchical structure.

Listing 2.14   Graphviz Graph Specifying Only Edges

```python
import graphviz as gv

# Create directed graph
G = gv.Digraph()

# Add edges and nodes
G.edges([('Satellite', 'Payload'),
         ('Satellite', 'Communication'),
```

```
                ('Satellite', 'Power'),
                ('Satellite', 'Propulsion'),
                ('Satellite', 'Attitude Control'),
                ('Satellite', 'Computer'),
                ('Communication', 'Antenna'),
                ('Communication', 'Transceiver'),
                ('Power', 'Solar Panels'),
                ('Power', 'Battery'),
                ('Propulsion', 'Thrusters'),
                ('Attitude Control', 'Reaction Wheels'),
                ('Attitude Control', 'Magnetorquers'),
                ('Payload', 'Camera')])

# Display graph
G
```

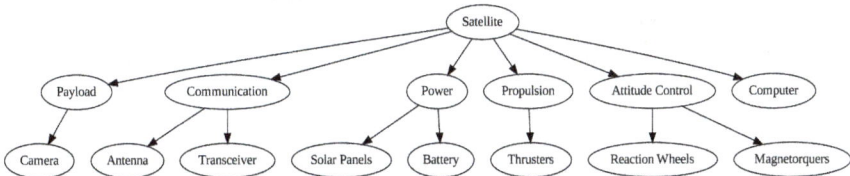

Figure 2.20   Satellite Component Tree

2.6.3   LAYOUT ENGINES

Graphviz provides several layout engines for positioning nodes and edges within a graph. Each engine uses a different algorithm to determine the placement of nodes in a two-dimensional space, and some allow for custom node placements. These layout engines produce visualizations with distinct structural properties. A full list of layout engines and their options is available at https://graphviz.org/docs/layouts/.

- dot is the default layout engine, used for directed graphs. It arranges nodes hierarchically, minimizing edge crossings.

- neato implements a spring model (force-directed) layout for undirected graphs. It minimizes a global energy function to produce visually balanced layouts, which are less structured compared to dot. Custom node placement is supported for greater control.

- fdp is similar to neato, but optimized for larger graphs, using a multi-scale approach to generate layouts more efficiently.

- sfdp is a multi-scale version of fdp, specifically designed for very large graphs with a focus on scalability. Custom node placement is supported for greater control.

- twopi arranges nodes in concentric circles, with one node at the center and other nodes placed in layers radiating outward. This layout is useful for visualizing radial structures or hierarchical data that fans out from a single central point.

- circo creates circular layouts, useful for visualizing graphs with cyclic structures or feedback loops.

- osage is designed for clustering and layered graphs. Unlike dot, it focuses on preserving node proximity within clusters rather than enforcing a strict hierarchy. Custom node placement is supported.

- patchwork generates a squarified treemap-like layout where nodes are represented as rectangles arranged in a hierarchical structure. This is useful for visualizing hierarchical data where node size represents importance or weight.

2.6.4 NETWORKX COMPATIBILITY

Graphviz uses identical network descriptions as NetworkX, making it straightforward to improve upon the limited visualization features of NetworkX. In this next example, Graphviz will be used to draw a translation of the NetworkX circuit board manufacturing network model subjected to computational analysis in Section 2.5.5. It is best visualized as a sequential process flow with parallel paths and operations. The script will convert the NetworkX graph object into Graphviz where the flow will be drawn in a left to right format with parallel nodes drawn at the same "rank" as computed with Graphviz, which isn't feasible with NetworkX.

In Listing 2.15, the function networkx_to_graphviz converts a NetworkX graph into a Graphviz graph, ready for visualization. It initializes a Digraph object with the specified svg format and default node attributes for a box shape. It then sets the rank direction to 'LR' (left-to-right) for the graph layout. Nodes and edges are added by iterating through the nodes and edges of the NetworkX graph to create Graphviz nodes with the same labels. It returns the created graph to be rendered as in Figure 2.21, as contrasted to the default NetworkX graph in Figure 2.16. Using the common NetworkX data structure, network computations are afforded with NetworkX as well as sophisticated visualization with Graphviz.

Listing 2.15   Converting NetworkX to Graphviz

```python
import networkx as nx
import graphviz as gv


def networkx_to_graphviz(graph):
    """Converts a NetworkX graph into a Graphviz Digraph.
```

```
Args:
  graph: A NetworkX graph object.

Returns:
  A Graphviz Digraph object.

This function creates a Graphviz representation of the given ↙
    NetworkX graph,
preserving node and edge relationships. The Graphviz graph is ↙
    configured with
a 'LR' layout and box-shaped nodes by default.
"""
dot = gv.Digraph(format='svg', node_attr={'shape': 'box'})
dot.attr(rankdir='LR')

for node in graph.nodes:
    dot.node(node, label=node)

for edge in graph.edges:
    dot.edge(edge[0], edge[1])

return dot

# Create a graph object
G = nx.DiGraph()

# Manufacturing process
# Add edges with duration attribute
G.add_edge("Board Stock", "Fabrication", duration=4)
G.add_edge("Fabrication", "Routing", duration=4)
G.add_edge("Routing", "Assembly 1", duration=4)
G.add_edge("Routing", "Assembly 2", duration=4)
G.add_edge("Assembly 1", "Test Station 1", duration=1)
G.add_edge("Assembly 1", "Test Station 2", duration=4)
G.add_edge("Assembly 2", "Test Station 1", duration=4)
G.add_edge("Assembly 2", "Test Station 2", duration=4)
G.add_edge('Test Station 1', 'Packaging', duration=2)
G.add_edge('Test Station 2', 'Packaging', duration=2)

# Convert to graphviz and display
networkx_to_graphviz(G)
```

Rendering of the shortest path is added in the next script in Listing 2.16. NetworkX is used to identify the path, which is sent to the `draw_shortest_path` function with the rest of the graph. The function converts the model to Graphviz while highlighting edges contained in the shortest path seen in the resulting Figure 2.22.

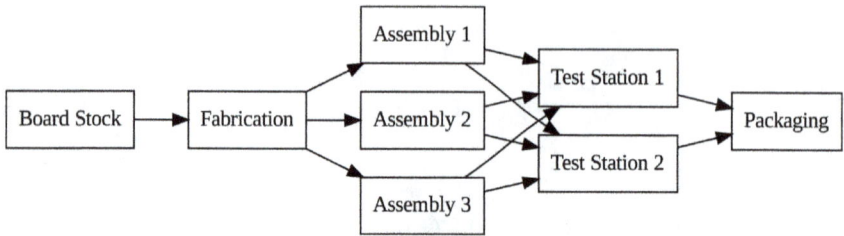

Figure 2.21   Circuit Board Manufacturing Process Flow with Graphviz

Listing 2.16   Drawing Shortest Path with Graphviz

```python
import networkx as nx
import graphviz as gv

def draw_shortest_path(graph, shortest_path):
    """Converts a NetworkX graph into a Graphviz Digraph, ↵
        highlighting the shortest path.

    Args:
        graph: A NetworkX graph object.
        shortest_path: A list of nodes representing the shortest path.

    Returns:
        A Graphviz Digraph object.
    """

    dot = gv.Digraph(format='svg', node_attr={'shape': 'box'})
    dot.attr(rankdir='LR', splines='line')

    for node in graph.nodes:
        dot.node(node, label=node)

    # Add edges, highlighting the shortest path
    for edge in graph.edges:
        if edge in adjacent_list(shortest_path):
            dot.edge(edge[0], edge[1], color='blue')
        else:
            dot.edge(edge[0], edge[1])

    return dot

def adjacent_list(input_list):
    """Creates an adjacent list of tuples from a given input list.
```

```
    Args:
        input_list: A list of elements.

    Returns:
        A list of tuples representing adjacent elements in the input ↙
            list.
    """

    adjacent_list = []
    for i in range(len(input_list) - 1):
        adjacent_list.append((input_list[i], input_list[i + 1]))
    return adjacent_list

# ... same manufacturing process model as above ...

shortest_path = nx.shortest_path(G, 'Board Stock', 'Packaging', ↙
    weight='duration')

# Draw graph highlighting shortest path
draw_shortest_path(G, shortest_path)
```

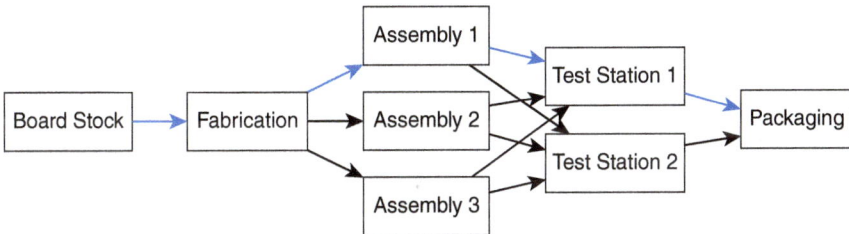

Figure 2.22   Circuit Board Manufacturing Process Shortest Path with Graphviz

2.6.5   CUSTOMIZATION

Graphviz has a wide variety of customizations for overall graph, node, and edge attributes. In conjunction with the underlying DOT language, graph clusters can be defined as groups of graphs. Edge relationships are possible within clusters, between clusters, or between nodes in separate clusters. Advanced graphic markup with HTML labels are possible. For more information see graph attributes at https://graphviz.org/docs/graph/, node attributes at https://graphviz.org/docs/nodes/, and edge attributes at https://graphviz.org/docs/edges/. The DOT language reference is at https://graphviz.org/doc/info/lang.html.

In this example, the LAN network diagram will be customized with Graphviz. It matches some of the custom styling from the NetworkX example in Section 2.18, but it also adds images for the nodes and explicitly sets the graph layout direction from left to right.

As shown in Listing 2.17, the graph object is created with `gv.Graph()` and customized with attributes for layout, node appearance, and edge properties. The `graph_attr` dictionary is used to set global attributes for the entire graph, such as the rank direction 'LR' for a left-to-right layout. The `node_attr` dictionary defines attributes for all nodes, where the shape is set to none to use images instead. The `edge_attr` dictionary specifies the edge width as 2 units.

When adding nodes with `G.node()`, the `image` attribute is used to specify the image file for each node, and the `labelloc='b'` option positions the label at the bottom of the node. The `G.edge()` calls define edge labels and set attributes including color and style. The resulting graph is shown in Figure 2.23.

Listing 2.17   Graphviz Customization

```python
import graphviz as gv

# Create graph with attributes
G = gv.Graph(graph_attr={'rankdir': 'LR'}, node_attr={'shape': ↙
    'none'}, edge_attr={'penwidth': '2'})

# Add nodes with images
G.node('Modem', image='modem.png', labelloc='b')
for device in ['Router1', 'Router2']:
    G.node(device, image='wireless-router.png', labelloc='b')

for node in ['Switch1', 'Switch2', 'Switch3']:
    G.node(node, image='switch.png', labelloc='b')

# Add edges
G.edge('Modem', 'Router1', label="coax")
G.edge('Router1', 'Router2', label="WiFi", color='green', ↙
    style='dashed')
G.edge('Router1', 'Switch1', label="ethernet", color='blue')
G.edge('Router2', 'Switch2', label="ethernet", color='blue')
G.edge('Router2', 'Switch3', label="ethernet", color='blue')

# Visualize the graph
G.view()
```

The next example in Listing 2.18 demonstrates a customized function for drawing a system context diagram of a GPS satellite showing interfaces to external systems. The `neato` layout engine is chosen that will place the system node in the center and arrange the surrounding nodes in a circular fashion. It takes care of wrapping text for very long labels and allows for custom labels with descriptive unicode

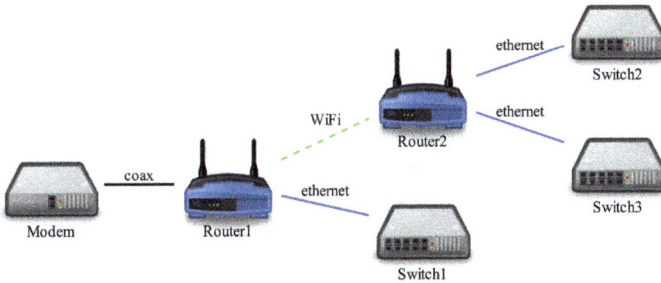

Figure 2.23   Customized LAN Network Graph with Graphviz

characters. The HTML labeling allows for fine control with the allowable tags at
https://graphviz.org/doc/info/shapes.html#html. The result is displayed
in Figure 2.24.

Listing 2.18   Context Diagram with Graphviz

```python
import graphviz
import textwrap

def context_diagram(system, external_systems, filename=None, ↙
    format='svg', engine='neato'):
    """
    Returns a context diagram with default boxes for nodes or allows ↙
        optional unicode labels without boxes.

    Parameters
    ----------
    system : tuple or string
        The name of the system to label the diagram. If it's a tuple, ↙
            the first element is the text label and the second is the ↙
            unicode icon.
    external_systems : list of tuples or strings
        Names of the external systems that interact with the system ↙
            in a list.
    filename : string, optional
        A filename for the output not including a filename extension. ↙
            The extension will be fspecified by the format parameter.
    format : string, optional
        The file format of the graphic output.

    Returns
    -------
    g : graph object view
```

```python
        Save the graph source code to file, and open the rendered ↙
            result in its default viewing application. PyML calls the ↙
            Graphviz API for this.
    """

    wrap_width = 12
    def wrap(text): return textwrap.fill(
        text, width=wrap_width, break_long_words=False)
    node_attr = {'color': 'black', 'fontsize': '11', 'fontname': ↙
        'arial',
                'shape': 'box'} # 'fontname': 'arial',
    c = graphviz.Graph('G', node_attr=node_attr,
                    filename=filename, format=format, engine=engine)
    if isinstance(system, tuple):
        # Write html label for unicode font size and label placement
        system_name = wrap(system[0])
        c.node(system_name, label=f'''<<font ↙
            point-size="30">{system[1]}</font><br/><font ↙
            point-size="11">{system[0]} </font>>''', labelloc="b", ↙
            shape='none')
    else:
        system_name = wrap(system)
        c.node(wrap(system_name))

    for external_system in external_systems:
        if isinstance(external_system, tuple):
            # Write html label for unicode font size and label placement
            c.node(wrap(external_system[0]), label=f'''<<font ↙
                point-size="30">{external_system[1]}</font><br/><font ↙
                point-size="11">{external_system[0]} </font>>''', ↙
                labelloc="b", shape='plain')
            c.edge(system_name, wrap(external_system[0]), len="1.4")
        else:
            c.edge(system_name, wrap(external_system), len="1.4")

    if filename is not None:
        c.render() # render and save dot file

    return c

# Unicode characters not seen in LaTeX listing
system = ('GPS Satellite', ' ')
actors_and_external_systems = [('airplane', ' '), ('ship', ' '), ↙
    ('heli', ' '), ('auto', ' '),
                        ('ground control', ' '),('other ↙
                            satellite', ' '), ('mobile device', ' ↙
                            ')]
context_diagram(system, actors_and_external_systems, ↙
    filename='satellite_context_diagram', format='svg')
```

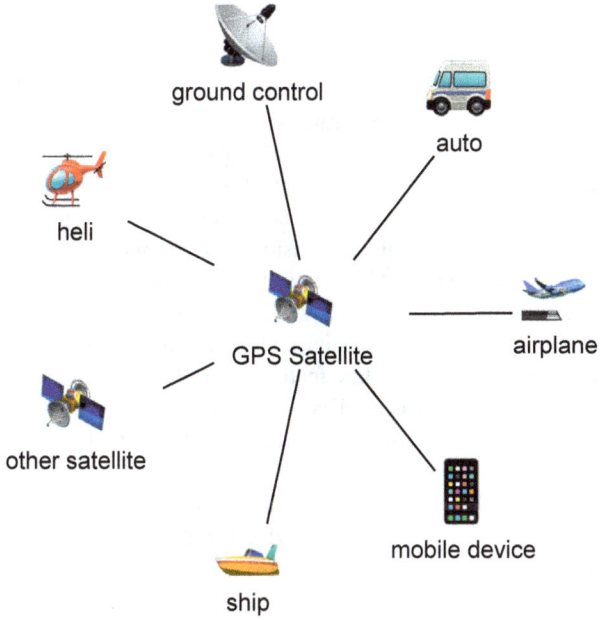

Figure 2.24   Satellite Context Diagram with Graphviz

2.7 SUMMARY

Using open-source Python libraries provides a wide range of benefits, including cost-effectiveness, reusability, community support, interoperability, flexibility, security, and innovation. These libraries integrate seamlessly with one another, forming a powerful ecosystem for scientific and engineering computations.

NumPy serves as the foundational library for efficient data manipulation, complex mathematical calculations, and large-scale scientific computing. It focuses on the efficient manipulation of multi-dimensional arrays with elements of the same type. NumPy offers a comprehensive suite of mathematical functions for operating on these arrays, streamlining data analysis, modeling complex systems, and solving engineering problems with precision and speed. Its emphasis on performance, particularly for large datasets, is a key feature. NumPy's ability to apply vectorized operations to entire arrays at once, eliminating the need for explicit loops, greatly accelerates computations. It is also tightly integrated with other libraries like Pandas, SciPy, and Matplotlib.

Pandas provides a robust framework for data manipulation and analysis, centered around data structures like the DataFrame, which is ideal for handling tabular data. It simplifies tasks such as data cleaning, transformation, and aggregation. Built on top of NumPy, Pandas excels at managing labeled data and offers extensive functionality for filtering, grouping, and analyzing large datasets. DataFrames can easily be created from files, databases, or NumPy arrays, making Pandas highly versatile for both small- and large-scale data processing.

SciPy extends the functionality of NumPy by providing a wide range of numerical algorithms and mathematical functions. These include modules for optimization, integration, linear algebra, and signal processing, allowing users to solve more complex mathematical problems. SciPy operates seamlessly with NumPy arrays and includes specialized sub-modules for various scientific computations.

Matplotlib is a comprehensive library for creating static, animated, and interactive visualizations. It supports a wide range of chart types, including line plots, histograms, and 3D plots. Built on top of NumPy, Matplotlib enables easy plotting of array data, making it particularly useful for visualizing results from simulations, experiments, or data analyses.

NetworkX offers extensive capabilities for network and graph analysis, making it invaluable for modeling systems. These models are critical for optimizing network performance, designing resilient systems, and identifying key components that may affect system functionality. The type of graph chosen depends on the nature of the system's relationships, which can involve undirected, directed, or bidirectional connections, and single or multiple edges between nodes.

Graphviz is a powerful general-purpose library for visualizing graphs described as collections of nodes connected by edges. It is particularly useful for visualizing networks, trees, circuits, and other structures common in engineering applications. Graphviz is compatible with NetworkX and enhances its limited visualization capabilities.

2.8 EXERCISES

1. Write a simulation that models the random movement of particles in a 2D space using the Monte Carlo method. Each particle should move a random distance in a random direction at each time step. Track the distance each particle moves from its origin after a large number of iterations. Visualize the results using Matplotlib.

2. Generate a finite element mesh for a 2D object using NumPy to model a chosen physical phenomenon (e.g., heat conduction, structural deformation, electrostatic potential, or fluid flow). Use NumPy to solve the system of linear equations derived from the finite element method, given a set of boundary conditions. Visualize the simulation results using Matplotlib.

3. Use NumPy's broadcasting feature to apply a convolution function to an image or an acoustic signal. Compare the original and processed results to demonstrate the effect of the convolution.

4. Read an audio signal from a file, perform a Fourier transform using NumPy, and identify the dominant frequency component. Use Matplotlib to plot the frequency spectrum and highlight the dominant frequency.

5. Model the dynamics of a simple mechanical system using its differential equation. Represent the system's transfer function using NumPy, and simulate the system's response to step and ramp inputs. Visualize the response using Matplotlib.

6. Formulate and solve an optimization problem for a chosen system. Some example problems include:

   - Minimizing the weight or material usage of a mechanical structure while maintaining strength and stiffness.

   - Minimizing the total cost of a manufacturing process, subject to constraints such as material usage and production time.

   - Minimizing energy consumption or power loss in a dynamic system while maintaining threshold levels for speed, precision, or power output.

   - Maximizing the strength or safety factor of a structure under load conditions, based on dimensions of structural elements or material selection.

   Use SciPy's `optimize.minimize()` function to find the optimal parameters that satisfy the constraints. Visualize the optimized function over different parameter values using Matplotlib. Perform a sensitivity analysis by perturbing input parameters and observing how the optimized solution changes.

7. Formulate an optimization problem for a static system where the objective is to optimize steady-state parameters (e.g., minimizing material usage or maximizing structural integrity in a design). Use SciPy's optimization functions

to solve for the optimal parameters that achieve the desired objective. Use Matplotlib to visualize the system's performance based on different parameter values, comparing the initial and optimized states.

8. Use Pandas to read an engineering dataset containing multiple parameter values over time. Plot selected parameter series from the DataFrame using Matplotlib by specifying them with the `data` parameter.

9. Read an image from a file and use NumPy to apply a Gaussian or other filter to the image. Use Matplotlib to display the original and filtered images side-by-side for comparison.

10. Model an electric circuit and solve for node voltages using nodal analysis. Use NumPy to construct the system of linear equations based on Kirchhoff's current law. Solve for node voltages using SciPy's linear algebra solvers. Simulate the circuit with different input voltages and resistances, and compare the solutions. Plot the voltages at various nodes in the circuit using Matplotlib.

11. Modify the projectile animation to include graphical user inputs and allow the animation to repeat. Have the user provide initial parameters before each animation run.

12. Create a continuous simulation framework using the Matplotlib animation module. At each time step, integrate the differential equation derivatives to calculate the updated state variables. Use the `init` function to specify initial conditions, and implement an integration routine (such as Euler's method, Runge-Kutta, or SciPy integration) in the update function. Plot the derivatives and integrals over time as the simulation progresses.

13. Generate artificial data to fit a specified exponential decay or another chosen function. Incorporate an appropriate degree of randomness to simulate physical behavior and data recording. Use SciPy's curve-fitting function to fit the data to the chosen model.

14. Using the NetworkX library, model and analyze a system of interconnected devices (e.g., IoT network, traffic network, or power grid). Define parameters such as node capacity, connectivity, or flow constraints, and simulate the network's performance under different failure or stress scenarios (e.g., node or link failures, traffic congestion, or overload). Solve for metrics such as network efficiency or reliability using appropriate NetworkX functions and visualize the network structure.

2.9   ADVANCED EXERCISES

1. Write a script to animate dynamic engineering phenomena for various scenarios. Enable saving the animation file for later viewing. Examples include:

- Fluid flow properties over time
- Material stress under dynamic loads
- Heat distribution in 2D or 3D space
- Traffic flow simulation

2. Create a discrete event simulation of a chosen system using Matplotlib's animation functionality. Events should be triggered according to an internal event calendar, and variables should be recalculated at each event. Display ongoing statistics as the simulation progresses.

3. Develop an optimization problem and solve it using SciPy's `optimize` module. Define a system with parameters to optimize in order to achieve a specific objective, such as minimizing error, maximizing efficiency, or reducing cost. Use Matplotlib to visualize the system's performance or behavior over time for different parameter values.

4. Build an engineering simulation using random sample inputs generated by NumPy that follow specified probability distributions (e.g., normal, exponential). Use Matplotlib to visualize the simulation. Employ Pandas to store and manipulate the data, and use SciPy to calculate key output statistics such as mean, variance, and confidence intervals for selected measures.

5. Define a dynamic system where the parameters need to be optimized over time (e.g., optimizing control parameters in a system with time-dependent behavior, such as minimizing energy consumption over time). Incorporate time-varying constraints and objectives into the optimization problem. Use SciPy's dynamic optimization functions, such as those in `scipy.optimize` or `scipy.integrate`, and visualize the system's behavior as the optimization progresses. Use Matplotlib to visualize the system's performance over time based on different parameter values.

6. Extend the exercise by using NetworkX to model and analyze a system of interconnected devices or components. Visualize the initial network structure and the results of failure simulations using Graphviz and Matplotlib. Generate graphical outputs to compare the network before and after failure events, highlighting critical paths and affected nodes. Use Matplotlib to plot relevant metrics (e.g., average path length, network connectivity) as a function of varying failure rates or stress levels.

7. Create an optimization problem where multiple objectives must be balanced (e.g., minimizing cost while maximizing efficiency in an industrial process). Use SciPy's `optimize` module to solve for Pareto optimal solutions, and visualize the trade-offs between competing objectives using Matplotlib.

8. Develop a Monte Carlo simulation using NumPy to analyze a system subject to uncertainty (e.g., financial risk, system reliability). Define random variables

based on specific probability distributions, and simulate the system's performance under varying conditions. Use Matplotlib to visualize the simulation results, and employ SciPy to calculate statistical measures such as confidence intervals and expected outcomes.

9. Design a simulation that combines both discrete event and continuous dynamic models (e.g., a manufacturing system with continuous flow and discrete process steps). Use `scipy.integrate` for the continuous dynamics and a discrete event simulator (such as SimPy or custom logic). Visualize the system's operation over time, highlighting how the discrete and continuous components of the system interact.

10. Use NetworkX and SciPy to model and optimize flow through a network (e.g., optimizing traffic flow or data transmission through a network of servers). Set up a multi-node network with capacity constraints, and use SciPy's optimization functions to find the optimal routing or flow distribution. Visualize the flow through the network and the impact of different optimization strategies using Graphviz.

3 Engineering Analysis Examples

This chapter highlights some engineering analysis examples building on previous chapters. There are innumerable engineering applications, so the few examples herein are intended to show a broad spectrum. They are purposely short for the sake of illustration and easy understanding on a page or two. They are starting points for further elaboration and scaling up for added complexity.

Each example will be introduced with an engineering purpose and relevant equations, where appropriate. Any new aspects or features beyond previous chapters are called out and explained. Otherwise the docstrings, comments, and code should speak for themselves and won't be duplicated in the overviews.

All the examples are demonstrated in a standard Python interpreter environment. The next chapter shows different forms that Python applications can take on for web-based and desktop platforms in other usage scenarios.

3.1 STATISTICAL ANALYSIS

The engineering applications of statistics and probability distributions are many to account for uncertainty. Per Chapter 2 there are core Python modules and other libraries with overlapping capabilities in these areas. This section shows a few examples of traditional parametric model building, confidence intervals, and hypothesis testing, generation of random numbers from probability distributions, and analysis of distribution data.

3.1.1 REGRESSION ANALYSIS

Regression analysis is a method to investigate relationships between variables for modeling phenomena. The degree of association between a dependent variable and one or more independent variables is measured statistically. These examples use the *statsmodel* library which supports estimation of statistical models, statistical testing, and data exploration. It is the most specific and comprehensive library for regression analysis, among other types of statistical models. SciPy also covers regression but is broader as is overviewed in Chapter 2. It is demonstrated for hypothesis testing in Section 3.1.4.

Single Independent Variable

The standard ordinary least squares (OLS) method for linear regression will be used, which minimizes the sum of squared residuals between actual observed values and predicted values of the dependent variable. The linear regression equation for a single independent variable is:

$$y = a + bx$$

Where:

y is the dependent variable
x is the independent variable
a is the intercept
b is the linear coefficient.

It is desired to test and quantify the dynamic properties of new materials under development. Measurements were collected on a new compound to determine the relationship between resistance and temperature. *statsmodel* will be used to solve for the coefficients and assess the fit and resulting statistics. In Listing 3.1 these are input as lists of X and Y data to *statsmodel* for an OLS model. In this case a non-zero intercept was specified since similar materials have resistance at zero temperature. Full details of the *statsmodels* OLS API are at [20].

The model is then subjected to a best fit method that returns a regression results object. A results summary is printed indicating a very strong relationship with an intercept of $a = 59.3$ and temperature coefficient $b = 1.3$.

Listing 3.1   Linear Regression with Inline Data

```python
import statsmodels.api as sm
import matplotlib.pyplot as plt


temperatures = [0, 50, 100, 150, 200, 250, 300, 350, 400]
resistances = [93, 122, 153, 248, 326, 363, 482, 518, 584]


X = temperatures
Y = resistances


# create OLS model with data
model = sm.OLS(Y, sm.add_constant(X)) # use (Y, X) for zero intercept


# best fit method
OLS_results = model.fit()
print(OLS_results.summary())
```

```
                           OLS Regression Results
==============================================================================
Dep. Variable:                      y   R-squared:                       0.984
Model:                            OLS   Adj. R-squared:                  0.982
Method:                 Least Squares   F-statistic:                     442.1
Date:                Thu, 25 Jan 2024   Prob (F-statistic):           1.38e-07
Time:                        21:07:59   Log-Likelihood:                -40.279
No. Observations:                   9   AIC:                             84.56
Df Residuals:                       7   BIC:                             84.95
Df Model:                           1
Covariance Type:            nonrobust
```

```
=========================================================================================
                 coef      std err           t      P>|t|        [0.025      0.975]
-----------------------------------------------------------------------------------------
const         59.3333       14.813       4.006      0.005        24.307      94.360
x1             1.3083        0.062      21.026      0.000         1.161       1.455
=========================================================================================
Omnibus:                     0.016   Durbin-Watson:                          1.990
Prob(Omnibus):               0.992   Jarque-Bera (JB):                       0.156
Skew:                        0.013   Prob(JB):                               0.925
Kurtosis:                    2.355   Cond. No.                               439.
=========================================================================================
```

Additional properties are available in the OLS regression results object. The following prints out primary statistics for the correlation coefficient R^2, f-statistic p-value, and p-values for the coefficients. The p-values can be subjected to hypothesis testing against given significance thresholds (see Section 3.1.4). It also prints out the coefficients and predicted values, which are attributes for the regression results object.

```python
print("R-squared: ", OLS_results.rsquared)
print("Prob (F-statistic):", OLS_results.f_pvalue)
print("p-values:", OLS_results.pvalues)
print("Coefficients: ", OLS_results.params)
print("Predicted Values:", OLS_results.predict(sm.add_constant(X)))
```

```
R-squared:  0.9844126608032431
Prob (F-statistic): 1.384471289860644e-07
p-values: [5.15302813e-03 1.38447129e-07]
Coefficients:  [59.33333333 1.30833333]
Predicted Values: [ 59.33333333 124.75       190.16666667 255.58333333 321.
 386.41666667 451.83333333 517.25       582.66666667]
```

The next script plots the results with the predicted relationship against the data points shown in Figure 3.1.

```python
# Best fit line
predicted = OLS_results.predict(sm.add_constant(X))

fig, ax = plt.subplots()
ax.scatter(X, Y, label='data points')
ax.plot(X, predicted, color='red', label='predicted')
ax.set(xlabel='Temperature (C)', ylabel= 'Resistance (Ohms)')
ax.legend()
```

In this example, the OLS results show the generic variable names y and x_1 since the input lists had no data labels. The next Listings 3.2 and 3.3 use Pandas dataframes with their richer data structure producing explicit variables names labeled on the output. This can help to understand and communicate the results especially with more independent variables.

Prolaunch desires to better estimate the testing effort for new launchers and software update releases. Data has been automatically collected on previous

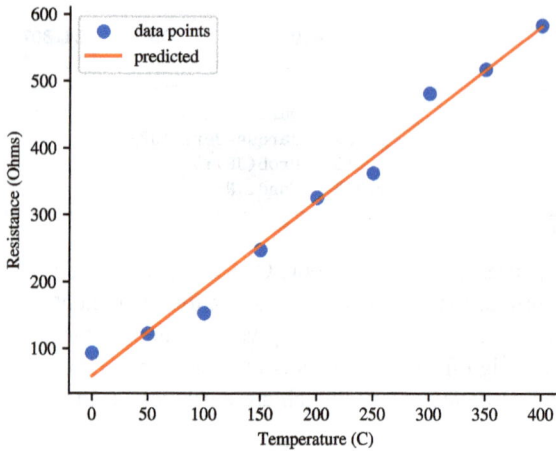

Figure 3.1   Linear Regression Results

developments on the counts of product features and actual testing effort. It is written to a csv file to be read by Pandas with the following structure:

```
Features,Testing Hours
14,129
25,277
9,122
...
```

Per Listing 3.2 that reads the csv file, the desired columns from the data frame are specified for the X and Y variables. This is sufficient to explicitly label the output results with the appropriate names.

Listing 3.2   Linear Regression with Data File

```python
import pandas as pd
import statsmodels.api as sm

# Read data file, specify dependent and independent variables
df = pd.read_csv ('features and testing hours.csv')
X = df [['Features']]
Y = df ['Testing Hours']

model = sm.OLS(Y, X) # zero intercept
OLS_result = model.fit()

print(OLS_result.summary())
```

```
                              OLS Regression Results
===============================================================================
Dep. Variable:              Testing Hours   R-squared (uncentered):           0.958
Model:                               OLS    Adj. R-squared (uncentered):      0.956
Method:                    Least Squares    F-statistic:                      434.2
Date:                   Fri, 02 Feb 2024    Prob (F-statistic):            1.51e-14
Time:                           20:27:13    Log-Likelihood:                 -99.744
No. Observations:                     20    AIC:                              201.5
Df Residuals:                         19    BIC:                              202.5
Df Model:                              1
Covariance Type:                nonrobust
===============================================================================
                 coef     std err         t      P>|t|     [0.025     0.975]
-------------------------------------------------------------------------------
Features      11.0135       0.529    20.836      0.000      9.907     12.120
===============================================================================
Omnibus:                           1.387    Durbin-Watson:                    2.531
Prob(Omnibus):                     0.500    Jarque-Bera (JB):                 1.070
Skew:                             -0.333    Prob(JB):                         0.586
Kurtosis:                          2.084    Cond. No.                         1.00
===============================================================================
```

The results show a robust model deriving the equation $hours = 11.0 * features$.

Multiple Independent Variables

Multivariable models are often necessary with several independent variables x_i to solve for the regression parameters in:

$$y = a + b_1 x_1 + b_2 x_2 + \ldots + b_n x_n$$

Custom drones are designed and built for a variety of applications, and a cost model is desired to estimate the development cost of future drones. Listing 3.3 reads data from previous drone projects for cost, drone weight, and data rate (as total sensor throughput rate) in a csv file with Pandas. The file has three columns with a header row of labels.

Two independent variables are specified for X in the OLS model creation. The results show a strong cost model with both independent variables of the form $cost = 113.3 * weight + 69.3 * data \ rate$.

A 3-dimensional scatter plot can be generated to show the regression line and visualization of the residuals against the data points. The script constructs a slightly customized plot resulting in Figure 3.2.

Listing 3.3   Multivariable Linear Regression and 3D Plot

```python
import pandas as pd
import statsmodels.api as sm

df = pd.read_csv("drone cost weight data rate.csv")
print(df)

X = df[['Weight (lbs)', 'Data Rate (Mbps)']]
```

```python
Y = df[['Cost']]

# apply OLS best fit to model
result = sm.OLS(Y, X).fit()
print(result.summary())

# Create figure
fig = plt.figure()
ax = plt.axes(projection ="3d")

# Scatter plot of actual data
ax.scatter3D(weight, data, cost, color = "green")

# Data for 3D regression line
weight_max = 120
data_max = 600

# Regression equation for max cost
cost_max = result.params[0] * weight_max + result.params[1] * data_max
weight_line = [0, weight_max]
data_line = [0, data_max]
cost_line = [0, cost_max]

# Plot regression line for data range
ax.plot3D(weight_line, data_line, cost_line, 'gray')
ax.set(xlabel='Weight (lbs)', ylabel="Data Rate (Mbps)")
ax.set_zlabel('Cost ($)', rotation=90, labelpad=1)
```

```
      Cost  Weight (lbs)  Data Rate (Mbps)
0    52167           101               640
1    29552            95               267
2    48166            73               511
3    21235            60               209
...
```

```
                           OLS Regression Results
==============================================================================
Dep. Variable:                   Cost   R-squared (uncentered):              0.974
Model:                            OLS   Adj. R-squared (uncentered):         0.970
Method:                 Least Squares   F-statistic:                         296.8
Date:                Fri, 02 Feb 2024   Prob (F-statistic):               2.25e-13
Time:                        20:39:36   Log-Likelihood:                    -183.07
No. Observations:                  18   AIC:                                 370.1
Df Residuals:                      16   BIC:                                 371.9
Df Model:                           2
Covariance Type:            nonrobust
==============================================================================
                     coef    std err          t      P>|t|      [0.025      0.975]
------------------------------------------------------------------------------
Weight (lbs)      113.2972     41.623      2.722      0.015      25.060     201.535
Data Rate (Mbps)   69.2747      8.168      8.481      0.000      51.959      86.591
==============================================================================
Omnibus:                       12.702   Durbin-Watson:                       1.771
Prob(Omnibus):                  0.002   Jarque-Bera (JB):                   10.009
Skew:                           1.478   Prob(JB):                          0.00671
Kurtosis:                       5.145   Cond. No.                             11.7
==============================================================================
```

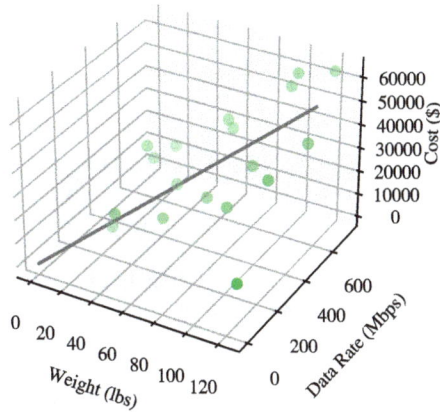

Figure 3.2   3D Scatter Plot and Regression

The examples in this section and other statistical models in statsmodel and SciPy are based on traditional regression methods. An alternative machine learning implementation for regression is shown in Section 3.11.

3.1.2 RANDOM DISTRIBUTIONS

The standard Python `random` library, NumPy, and SciPy provide functions for generating random numbers from many different probability distributions which can be used in simulations or other engineering applications. Single values are generated with the standard `random` library. NumPy and SciPy can generate arrays which are more suitable for large data and advanced applications. The standard library has basic distributions, NumPy has an extensive variety, while SciPy has even more available.

For example, the `random.random` function generates a random number uniformly distributed between 0 and 1 where all values are equally likely to occur. The `np.random.uniform(min, max)` function can generate an array of uniformly distributed random numbers in a given range, `np.random.normal()` generates random numbers from a normal distribution, and `np.random.exponential()` function can be used to generate random numbers from an exponential distribution.

NumPy distribution functions can generate a single random value or an array of values. By default a single value will be drawn and returned, while passing the optional `size` parameter will specify the number of values to generate and return in an array of that size. For example, `np.random.normal(size=5000)` will generate an array of 5000 normally distributed random values.

Random distribution examples using the standard `random` library for uniform and normal distributions are in Listing 3.4. NumPy examples with arrays containing uniform, normal, triangular, and Rayleigh distributions used in later examples are in Listing 3.5.

Listing 3.4   Random Number Generation

```python
import random
# Random number documentation at ↙
    https://docs.python.org/3.9/library/random.html


# Generate random number uniformly distributed between 0 and 1
random_uniform = random.random()
print(f"{uniform_number = }")


# Generate random integer uniformly distributed 100 and 200
random_integer = random.randint(100, 200)
print (f"{random_integer = }")


# Generate normal variate with mean 100 and standard deviation 15
random_normal = random.normalvariate(100, 15)
print(f"{random_normal = :6.2f}")

uniform_number = 0.5988667056940171
random_integer = 146
random_normal =  95.72
```

Listing 3.5   Random Number Generation with NumPy

```python
import numpy as np


# Generate 100 targeting errors from a normal distribution with mean ↙
    0 and standard deviation 2
targeting_errors = np.random.normal(0, 2, size=100) # degrees
print(f"{targeting_errors = } \n")


# Generate 1000 circuit board interarrival times using an ↙
    exponential distribution with a rate parameter of 0.5
interarrival_times = np.random.exponential(1/0.5, size=100) # minutes
print(f"{interarrival_times = }")

targeting_errors = array([ 3.54697887, -4.7699865 ,  0.74219808, ...
       -3.76256205, -1.32151273, -1.05908763, -0.37733315, -0.60752082,
       -2.56198889, -1.79604971,  3.89666748,  2.618721  , -0.80547759,
...
       -0.44212275, -3.01326576,  1.71312262, -0.76547141, -0.5576573 ])

interarrival_times = array([2.05942688, 3.50784738, 2.83070506, ...
       0.63299563, 4.780682  , 0.19634971, 1.17990081, 1.01361641,
       5.18943653, 3.02250791, 0.48280875, 2.63509959, 2.25860752,
...
       3.74872537, 2.82018547, 8.10153673, 0.0728005 , 1.87402718 ])
```

NumPy distributions can be directly plotted with Matplotlib. The following example in Listing 3.6 demonstrates a customized set of subplots from generated NumPy distributions. The same method would be used for the probabilistic outputs of an analysis.

Random data is generated using NumPy's `random` functions and then plotted as a distribution with `hist()`. A dictionary of keywords and arguments (*kwargs*) is defined to style the histograms for color and relative bin width spacing. The `**` syntax is used to pass the *kwargs* dictionary to the `hist()` function, which applies the styling parameters to all the histograms.

Listing 3.6   NumPy Random Distribution Histograms

```python
import numpy as np
import matplotlib, matplotlib.pyplot as plt
matplotlib.rcParams['axes.spines.top'] = False
matplotlib.rcParams['axes.spines.right'] = False

number_samples = 10000
num_bins = 12

# Populate numpy arrays with distributions
waiting_time = np.random.uniform(2,5,number_samples)
heading_error = np.random.standard_normal(number_samples)
payload_weight = np.random.triangular(40, 50, 70, number_samples)
wind_velocity = np.random.rayleigh(10, number_samples)

# Dictionary for default histogram keyword parameters
kwargs = {'color': "blue", 'rwidth': 0.9}

figure, ((axis1, axis2), (axis3, axis4)) = plt.subplots(2, 2)

# Create histograms of the distributions
axis1.set_title('Shuttle Waiting Time\n Uniform Distribution')
axis1.hist (waiting_time, num_bins, **kwargs)
axis1.set(xlabel = 'Minutes', ylabel ='Frequency')

axis2.set_title('Heading Error\n Normal Distribution')
axis2.hist (heading_error, num_bins, **kwargs)
axis2.set(xlabel = 'Degrees', ylabel ='Frequency')

axis3.set_title('Payload Weight\n Triangular Distribution')
axis3.hist (payload_weight , num_bins, **kwargs)
axis3.set(xlabel = 'Pounds', ylabel ='Frequency')

axis4.set_title('Wind Velocity\n Rayleigh Distribution')
axis4.hist (wind_velocity, num_bins, **kwargs)
axis4.set(xlabel = 'MPH', ylabel ='Frequency')

plt.subplots_adjust(hspace=1, wspace=0.5)
```

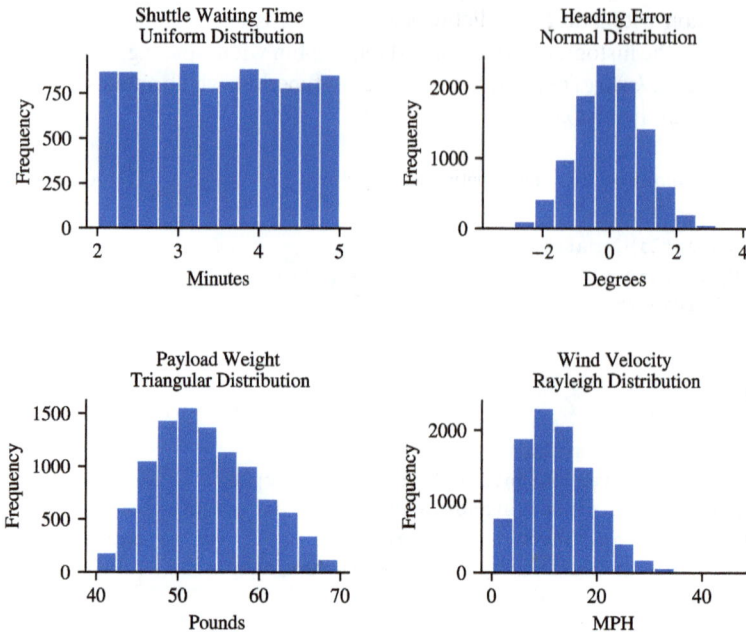

Figure 3.3   Histograms with NumPy and Matplotlib

3.1.3   EMPIRICAL DISTRIBUTION AND INVERSE TRANSFORM

A new lithium battery is being tested with accelerated cycles of charging and discharging to determine how many cycles until failure, when it is unable to hold a charge anymore. It is desired to use an empirical distribution from the test data to drive simulations of field usage. Empirical probability distributions are constructed directly from data as opposed to theoretical distributions, which may not be good representations of actual values. Listing 3.7 generates a normalized cumulative distribution and plots it as shown in Figure 3.4. NumPy functions are used to sort the data by value and arrange the continuous distribution.

Listing 3.7   Empirical Cumulative Distribution

```
import matplotlib.pyplot as plt
import numpy as np

# Cycles to failure
cycles = [5224, 2700, 939, 2054, 6926, 880, 3522, 4109, 1056, 1115, 1
    643, 1115]

# Sort data
sorted_data = np.sort(cycles)
```

```
# Create cumulative count array of evenly spaced values length of ↙
    the sorted data [0, 1, 2, ...] and normalize it 0-1
cum_probabilities = ↙
    np.arange(len(sorted_data))/float(len(sorted_data)-1)

# CDF
fig, axis = plt.subplots()
axis.set(xlabel='# Cycles to Failure',
        ylabel='Cumulative Probability',
        yticks=np.linspace(0, 1.0, 11)
        )
plt.grid(True)
axis.plot(sorted_data, cum_probabilities)
```

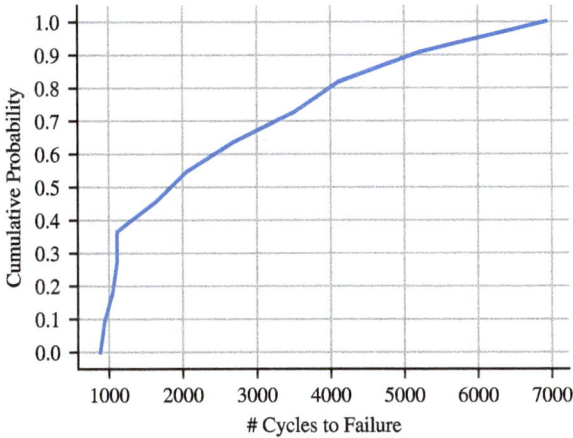

Figure 3.4   Empirical Cumulative Distribution for Battery Cycles to Failure

The CDF can be used for the inverse transform method by indexing it with a random number r that is uniformly distributed between 0 and 1, denoted as $U(0, 1)$. It is set equal to the cumulative distribution, $F(x) = r$, and x is solved for. For Monte Carlo analysis, a particular value r_i gives a value x_i, which is a particular sample value of X per:

$$x_i = F^{-1}(r_i)$$

This will generate values x_i of the random variable X with a distribution of values matching the empirical CDF. With enough samples, a uniform distribution for r will produce an X distribution that matches the empirical CDF for number of cycles until failure.

```
# inverse transform
# generate random values of cycles using empirical CDF with r_i = ↙
    U(0, 1)

for battery in range(1000):
    r = np.random.uniform()
    # linear interpolation
    cycles = np.interp(r, cum_probabilities, sorted_data)
    print(cycles)
1027
4363
2765
. . .
```

3.1.4 HYPOTHESIS TESTING

FastCircuit has a production line goal of 80% average yield of circuit cards on first pass. It measures the percent of cards started that pass all tests first time without defects or rework. Yield data has been collected on a small sample of production runs to be subjected to hypothesis testing for the entire population.

Hypothesis testing is used when it is desirable to know if the mean of a variable is equal to, less than, or greater than a specific value; or if there is a significant difference between two means. It is based on the assumption of normality of sample means. A null hypothesis, H_0, states a certain relationship that may or may not be true at a given significance level, α, which is the probability of rejecting a hypothesis when it is true. The hypothesis is not rejected if the relationship holds true statistically.

Hypothesis testing is employed with the `scipy.stats` module in Listing 3.8. They null hypothesis H_0 states that the average is less than the goal of 80%. A standard α of 5% is specified. The output indicates to reject the hypothesis and accept that the average is equal to or greater than the goal.

Listing 3.8 Hypothesis Testing

```
import scipy.stats as stats

# Define the null and alternative hypotheses for 1-sided test
yield_goal = 80
HO = f"mean yield <= {yield_goal}"
HA = f"mean yield > {yield_goal}"

# Significance level
alpha = .05

# Circuit card yield data
yields = [80.6, 99.0, 85.3, 83.8, 69.9, 89.6, 91.1 , 66.2, 91.2, 82.7↙
    , 73.5, 82.0, 54.0, 82.9, 75.9, 98.3, 107.2, 85.5, 79.1 , 84.3, 8↙
    9.3, 86.3, 79.0, 92.3, 87.0]
```

```
# Perform 1-sided t-test
t_statistic, p_value = stats.ttest_1samp(yields, yield_goal, ⤶
    alternative = "greater")

# Print the results
print("t-statistic:", t_statistic)
print("p-value:", p_value)

if(p_value < alpha):
  print("Reject null hypothesis and accept HA: " + HA)
else:
  print("Fail to reject null hypothesis H0: " + H0)
```

```
t-statistic: 1.7542295178350589
p-value: 0.04607666303535189
Reject null hypothesis and accept HA: mean yield > 80
```

3.1.5   CHI SQUARE TEST

A simulation model of airport traffic requires a probabilistic distribution for the inter-arrival times of incoming flights. A Chi Square goodness-of-fit test will be used to compare a theoretical distribution to the collected inter-arrival data below. The test measures whether the observed data is significantly different from the proposed distribution by dividing the data into intervals, and taking the difference between the number of occurrences in the observed data and the number expected by the proposed distribution.

Interval (Min.)	Observed Count
[0-10)	151
[10-20)	89
[20-30)	72
[30-40)	31
[40-50)	19
[50-60)	13
[60+]	9

An exponential distribution with a mean of 20 minutes will be tested as the theoretical distribution. A Chi Square test is performed with SciPy in Listing 3.9. The observed counts are provided with edges of the desired intervals. The loop calculates the theoretical probabilities from the SciPY exponential CDF function for comparison. The program then tests the null hypothesis that there is no significant difference between the observed and expected frequencies. The test result based on the p-value fails to reject the null hypothesis, indicating the exponential distribution is a good fit to the data.

Listing 3.9   Chi Square Test

```python
# Perform a Chi Square test on inter-arrival times against an ⤶
    exponential distribution with a mean of 20.

import scipy.stats as stats
from scipy.stats import chisquare
import numpy as np

# Use scipy's expon.cdf to compute the CDF at a given value x
def scipy_cdf(x, mean):
    return stats.expon.cdf(x, scale=mean) # scale is 1/lambda (mean)

# Observed frequencies
observed_counts = [151, 89, 72, 31, 19, 13, 9]
N = sum(observed_counts) # Total number of observations
# float('inf') ensures 100% coverage of the asympotic CDF
interval_edges = [0, 10, 20, 30, 40, 50, 60, float('inf')]

# Theoretical distribution mean and expected counts to populate
mean = 20
expected_counts = []

for i in range(len(interval_edges) - 1):
    lower_bound = interval_edges[i]
    upper_bound = interval_edges[i + 1]
    # Interval probability is difference between maximum and minimum ⤶
        CDF values
    interval_probability = scipy_cdf(upper_bound, mean) - ⤶
        scipy_cdf(lower_bound, mean)
    expected_counts.append(interval_probability * N)

X2, p_value = chisquare(observed_counts, expected_counts)
print(f'{X2=} {p_value=}')

# Hypothesis testing
# H0: there is no significant difference between the observed and ⤶
    expected frequencies.
alpha = .05 # significance level
if p_value < alpha:
    print("Reject the null hypothesis")
else:
    print("Fail to reject the null hypothesis")
X2=10.629398914521413 p_value=0.10052814836317078
Fail to reject the null hypothesis
```

3.2   PROJECTILE MOTION WITH AIR RESISTANCE

More realistic flight models are necessary to accurately design, calibrate, and use the ProLaunch systems. Previous calculations for projectile motion ignored the effects of air resistance causing drag, as if in operating in a vacuum. The updated equations for projectile horizontal and vertical motion with the effects of air resistance drag are:

$$x_{\text{drag}}(t) = v_0 \cos(\theta)t - \frac{1}{2}C_d \rho A v_0^2$$

$$h_{\text{drag}}(t) = v_0 \sin(\theta)t - \frac{1}{2}gt^2 - \frac{1}{2}C_d \rho A v_0^2 t/m$$

where:
- $x_{\text{drag}}(t)$ is the horizontal position with air resistance at time t
- $h_{\text{drag}}(t)$ is the vertical position with air resistance at time t
- C_d is the drag coefficient
- ρ is the air density (kg/m^3)
- A is the projectile cross-sectional area (m^2)
- m is the projectile mass (kg).

The equations are implemented in Listing 3.10. The generalized `trajectory` function takes projectile and drag parameters with a flag to incorporate drag. It simulates a trajectory over time until the projectile hits ground. It does not pre-compute the final time, so random effects can be incorporated in the time loop. The impact of drag for the given projectile is visualized in Figure 3.5.

Listing 3.10   Projectile Motion with Air Resistance Drag

```
import math
import matplotlib.pyplot as plt

# Constants
g = 9.81 # m/s^2

# Initial conditions
v0 = 32 # m/s
theta = 50 * math.pi / 180 # radians
height = 0

# Drag parameters
cd = 0.25 # drag coefficient
rho = 1.225 # air density (kg/m^3)

# Projectile parameters
A = 0.2 # projectile cross-sectional area (m^2)
m = 10 # projectile mass (kg)
```

```python
def trajectory(v0, theta, height, use_drag, cd, rho, A, m):
    """
    Computes the trajectory of a projectile with or without air drag.

    Parameters:
    v0 (float): Initial velocity (m/s)
    theta (float): Launch angle (radians)
    height (float): Initial height (m)
    cd (float): Drag coefficient
    rho (float): Air density (kg/m^3)
    A (float): Cross-sectional area of the projectile (m^2)
    m (float): Mass of the projectile (kg)

    Returns:
    t (list): Time points (s)
    x (list): Horizontal distances (m)
    h (list): Heights (m)
    """
    # Data output lists
    t, h, x = [], [], []

    time = 0
    dt= .1
    while height >= 0:
        height = v0* math.sin(theta) * time - 0.5* g * time**2 - 0.5* \
            cd * rho * A * v0**2 * time / m # gravity - drag
        distance = v0* math.cos(theta) * time - 0.5* cd * rho * A * v0 \
            **2 * time / m
        t.append(time)
        h.append(height)
        x.append(distance)
        time += dt
    return (t, x, h)

fig, axis = plt.subplots(figsize=(5,4))
axis.set(xlabel='Distance (m)', ylabel='Height (m)')

use_drag = True
_, x, h = trajectory(v0, theta, height, use_drag, cd, rho, A, m)
axis.plot(x, h, label="With drag")

use_drag = False
_, x, h = trajectory(v0, theta, height, use_drag, cd, rho, A, m)
axis.plot(x, h, label="No drag")

plt.legend()
```

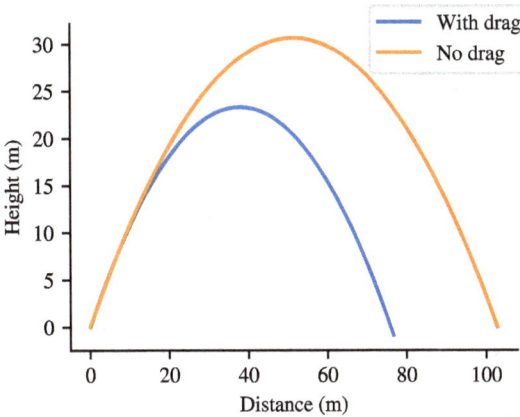

Figure 3.5   Projectile Motion with and without Drag

3.3   PRESENT VALUE

FastCircuit has economic decision scenarios to decide between investment options using present value analysis. Present Value (PV) is the value of a future cash flow discounted to the present to account for the time value of money. It reflects the current value of future money as if it existed today. Normalizing cash flows this way enables economic decision making with respect to today's value of projected cash flows.

Since an option is to invest money now and gain interest, a dollar today will be worth more in the future. If FastCircuit invests \$10,000 now into an interest bearing account of 6% compounded annually, the Future Value (FV) after 5 years would be $10000 * (1 + .06)^5 = \$13382$. The present value calculation reverses this per:

$$PV = FV/(1+r)^n$$

where:

PV is the present value of a cash flow
FV is the future value of a cash flow
r is the discount rate
n is the period (time step) for compounding.

A simple function for present value is:

```
def present_value(cash_flow, rate, n):
    """Calculates the present value of a cash flow for given rate ⤾
        and time period n."""
    return cash_flow / (1 + rate)**n
```

Net present value (*NPV*) calculates the cumulative difference between the present value of cash inflows and outflows over time. It provides a consistent framework to reconcile flows at different time points into a common unit *PV*. The net profit or loss normalized to current terms is useful to decide going forward on an engineering project or investment option. The equation for the net present value *NPV* of a series of *N* cash flows is:

$$NPV = \sum_{n=0}^{N} \frac{FV_n}{(1+r)^n}$$

FastCircuit is considering the purchase vs. rent of a photolithography machine for semiconductor fabrication over 3 years. It would cost $80,000 up front and could be resold for $35,000. Alternatively it could be rented for $15,000 per year with equal net future amounts.

The *NPV* calculation in the next script is used to compare the options in *PV* terms. Outflows are expressed as negative *FV* amounts. Initial investments at time zero incur no discount and are represented in the initial value (corresponding to the 0th value in the cash flow list). This also allows for scenarios with compounding at period beginnings. With the given interest rate of 8%, it is best to rent the system instead of tying up the upfront purchase cost.

```python
import numpy as np

def present_value(cash_flow, rate, n):
    """Calculates the present value of a cash flow for given rate ⤶
        and time period n."""
    return cash_flow / (1 + rate)**n

def net_present_value(cash_flows, rate):
    """Calculate net present value of cashflows"""
    """
    Args:
    cash_flows: A list of cash flows for each period..
    rate: The discount rate per period as a fraction.

    Returns:
    Net present values
    """
    present_values = [present_value(cash_flow, rate, time) for time, ⤶
        cash_flow in enumerate(cash_flows)]
    cumulative_npvs = list(np.cumsum(present_values))

    # return final NPV
    return cumulative_npvs[-1]

# Purchase photolithography machine for $80000 and resell for $35000 ⤶
    after 3 years
```

```
NPV_purchase = net_present_value(cash_flows=[-80000, 0, 0, 40000], ↙
    rate=.08)

# Rent machine for 3 years at $15000 per year
NPV_rent = net_present_value(cash_flows=[-15000, -15000, -15000], ↙
    rate=.08)

print(f'{NPV_purchase=:.0f}\n{NPV_rent=:.0f}')
NPV_purchase=-48247
NPV_rent=-41749
```

FastCircuit is also assessing options to improve their circuit card inspection and refurbishing operations. A laser measurement system is under consideration for automating circuit card identification, sorting, refurbishing, and repair. It can replace manual labor in the initial sorting of circuit cards placed on a moving conveyor.

The system will cost $50K and require $40K of labor costs to implement machine learning with it. It is expected to affect labor savings linearly increasing from $30K to $45K in the 4th year with training set improvements.

The next updated program in Listing 3.6 is more generalized and adds graphics output. It includes the provision to calculate effective interest rates via compounding given a nominal annual interest rate and an arbitrary number of periods. Graphic output is shown in Figure 3.6.

Listing 3.11   Net Present Value Analysis

```python
import matplotlib.pyplot as plt
from matplotlib.ticker import MaxNLocator, MultipleLocator

import matplotlib.pyplot as plt
from matplotlib.ticker import MaxNLocator

def present_value(cash_flow, rate, n):
    """Calculates the present value of a cash flow for given rate ↙
        and time period n."""
    return cash_flow / (1 + rate)**n

def net_present_value(cash_flows, annual_rate, n=12, plot=True, ↙
    plot_pv=False, show_values=False):
    """Calculate net present value of cashflows and optionally plot ↙
        them."""
    """
    Args:
    cash_flows: A list of cash flows.
    annual_rate: The annual nominal discount rate.
    n: Compounding frequency per year (default is 12) (optional)
    plot: Flag to display a bar plot of future values and line for ↙
        net present values (default is True) (optional)
    plot_pv: Flag to display present values on bar plot (default is ↙
        True) (optional)
```

```python
    show_values: Flag to display data values on the bars and line ↙
        (default is False) (optional)

    Returns:
    The cash flow present values and net present values for each ↙
        period.
    """
    effective_rate = annual_rate/n
    present_values = [present_value(cash_flow, effective_rate, time) ↙
        for time, cash_flow in enumerate(cash_flows)]
    cumulative_npvs = [sum(present_values[:i+1]) for i in ↙
        range(len(present_values))]

    if plot:
        plot_cashflows(cash_flows, present_values, cumulative_npvs, ↙
            plot_pv, show_values)

    return present_values, cumulative_npvs

def plot_cashflows(cash_flows, present_values, cumulative_npvs, ↙
    plot_pv=False, show_values=False):
    """Plot the future cashflows, their present values, and ↙
        cumulative NPV on a single axis."""

    # Separate positive (inflows) and negative (outflows) cash flows
    inflows = [value if value >= 0 else 0 for value in cash_flows]
    outflows = [value if value < 0 else 0 for value in cash_flows]

    # Create the figure and axis objects
    fig, ax = plt.subplots()
    bar_width = 0.25 # Set bar width to 1/4 of a division

    # Bar chart for inflows (green) and outflows (red)
    inflow_bars = ax.bar([i - bar_width / 2 for i in ↙
        range(len(inflows))], inflows, color='green', ↙
        width=bar_width, label='Inflow')
    outflow_bars = ax.bar([i - bar_width / 2 for i in ↙
        range(len(outflows))], outflows, color='red', ↙
        width=bar_width, label='Outflow')

    # Display optional data values at the top of each bar
    if show_values:
        for bar in inflow_bars + outflow_bars:
            ax.text(bar.get_x() + bar.get_width() / 2, ↙
                bar.get_height(), f"{bar.get_height():.1f}", ↙
                    ha='center', va='bottom' if bar.get_height() >= 0 ↙
                        else 'top')

    # Bar chart for adjusted present value flows (if plot_pv is True)
```

```python
    if plot_pv:
        pv_bars = ax.bar([i + bar_width / 2 for i in ↵
            range(len(present_values))], present_values, ↵
            color='purple', width=bar_width, label='Present Value')
        if show_values:
            for bar in pv_bars:
                ax.text(bar.get_x() + bar.get_width() / 2, ↵
                    bar.get_height(), f"{bar.get_height():.1f}",
                        ha='center', va='bottom' if bar.get_height() ↵
                            >= 0 else 'top')

    # Line plot for cumulative NPV
    npv_line = ax.plot(range(len(cash_flows)), cumulative_npvs, ↵
        color='blue', marker='o', linestyle='dashed', linewidth=2, ↵
        markersize=5, label='Cumulative NPV')

    # Add data labels for NPV line points if show_values is True
    if show_values:
        for i, npv in enumerate(cumulative_npvs):
            ax.text(i, npv, f"{npv:.1f}", ha='center', va='bottom' if ↵
                npv >= 0 else 'top')

    # Titles and labels
    ax.set(title='Cash Flows and Cumulative NPV', xlabel='Time ↵
        Period', ylabel='Value ($K)')
    ax.axhline(0, color='black', linewidth=0.5)
    ax.legend()

    # Ensure that x-axis only shows integer values for periods
    ax.xaxis.set_major_locator(MaxNLocator(integer=True))

    # Adjust layout to prevent clipping of labels
    plt.tight_layout()

# Laser inspection machine costs and savings
cash_flows = [-50-40, 30, 35, 40, 45] # $K
annual_rate = 0.07

net_present_value(cash_flows, annual_rate, n=1, plot=True, ↵
    plot_pv=True, show_values=False)
```

The positive NPV in this example justifies the purchase. The laser measurement system is the backdrop for machine learning in Section 3.11.2 for identification of electronic circuit cards.

This script is also contained in a web application in Section 4.8. It is the subject of automated testing described in Section 5.4.1. It was iteratively developed with a testing script that was run after each change. This ensured it passed test cases by comparing the outputs with expected results without manual testing effort.

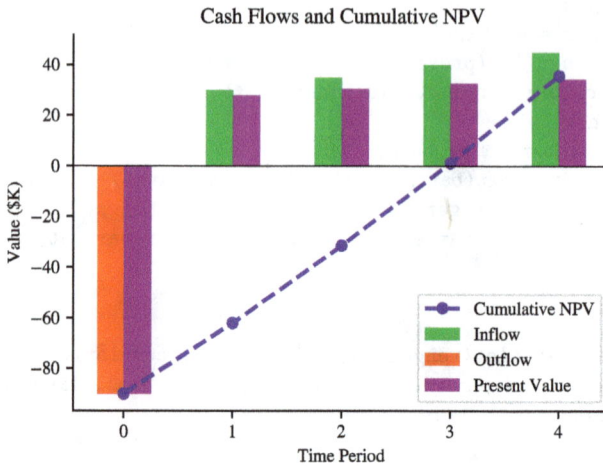

Figure 3.6   Net Present Value Analysis

3.4   PROJECT TASK NETWORK CRITICAL PATH

FastCircuit is planning the development of a new circuit board with GPS and wants to know the minimum time to complete the project and which activities are critical to complete in that time. Critical path analysis will be used to determine the time and path from a task network of connected activities with individual time durations.

Listing 3.12 uses a project task network defined in a dictionary, NetworkX to generate the network structure and determine the paths, an algorithm to find the critical path, and Graphviz to arrange and render the visualization. The resulting task network and critical path is in Figure 3.7.

NetworkX has ample facilities for computing with edge data. However, it doesn't provide any functions for quantitative node attributes so this implementation was developed using the `duration` node attribute. The function `find_max_path_key` finds the maximum value from a newly created dictionary of path data. The task dictionary tree structure unpacking replaces numerous hand coded NetworkX and Graphviz function statements for each node and edge and the resulting redundancies (e.g. single nodes with many successors would require many edge statements).

Listing 3.12   Critical Path Analysis with NetworkX and Graphviz

```
import NetworkX as nx
import graphviz as gv
import textwrap

wrap_width = 12
def wrap(text): return textwrap.fill(
    text, width=wrap_width, break_long_words=False)
```

```python
def draw_task_network(graph, critical_path_edges):
    dot = gv.Digraph(format='pdf', engine='dot', node_attr={'shape': ↵
        'square', 'width': '.3', 'fontsize': '10', 'fontname': ↵
        'times'})
    dot.attr(rankdir='LR')

    for node in graph.nodes:
        dot.node(node, label=wrap(node) + ↵
            "\n\n"+str(graph.nodes[node]['duration'])+" days")

    for edge in graph.edges:
        if edge in critical_path_edges:
            dot.edge(edge[0], edge[1], color='red')
        else:
            dot.edge(edge[0], edge[1])

    return dot

# Convert project tasks into NetworkX graph
def tasks_to_graph(project_tasks):
    G = nx.DiGraph()
    for node, attrs in project_tasks.items():
        G.add_node(node, duration=attrs['duration'])
        for successor in attrs['successors']:
            G.add_edge(node, successor)
    return G

def find_max_path_key(dictionary):
  """Finds the dictionary key for the maximum value in dictionary"""
  max_value = max(dictionary.values())
  max_path_key = None
  for key, value in dictionary.items():
    if value == max_value:
      max_path_key = key

  return max_path_key

def create_edges_from_nodes(nodes):
  """Returns a list of edges from a tuple of nodes."""
  edges = []
  for i in range(len(nodes) - 1):
    edges.append((nodes[i], nodes[i + 1]))

  return edges

# Project task dictionary of dictionaries
project_tasks = {
    'fabrication': {'duration': 10, 'successors': ['circuit board ↵
```

```python
          assembly']},
    'final testing': {'duration': 5, 'successors': []},
    'system requirements': {'duration': 15, 'successors': ['chip ↙
        design', 'firmware development', 'GPS driver software ↙
        development', 'test data']},
    'chip design': {'duration': 22, 'successors': ['fabrication']},
    'circuit board assembly': {'duration': 4, 'successors': ↙
        ['breadboard testing']},
    'GPS driver software development': {'duration': 28, ↙
        'successors': ['breadboard testing']},
    'test data': {'duration': 12, 'successors': ['testing hardware']},
    'testing hardware': {'duration': 18, 'successors': ['breadboard ↙
        testing']},
    'firmware development': {'duration': 15, 'successors': ↙
        ['breadboard testing']},
    'card production': {'duration': 10, 'successors': ['final ↙
        testing']},
    'breadboard testing': {'duration': 3, 'successors': ['redesign', ↙
        'software updates']},
    'software updates': {'duration': 10, 'successors': ['final ↙
        testing']},
    'redesign': {'duration': 12, 'successors': ['card production']}
}

# Convert the dictionary back to a NetworkX graph
graph = tasks_to_graph(project_tasks)
graph.nodes(data=True), graph.edges()

# Find all paths from beginning to end
paths = nx.all_simple_paths(graph, 'system requirements', 'final ↙
    testing')

# Dictionary comprehension with path tuples as keys
# Values are sums of task durations for paths
path_durations = {tuple(path): sum(graph.nodes[node]['duration'] for ↙
    node in path) for path in paths}

# Determine critical path
max_path_key = find_max_path_key(path_durations)
critical_path_edges = create_edges_from_nodes(max_path_key)
print(f"The maximum path is {max_path_key}")
print(f'{critical_path_edges=}')

# Draw critical path network graph
g = draw_task_network(graph, critical_path_edges)
g.render(filename='project_network_critical_path')

The critical path is ('system requirements', 'chip design', ... )
The minimum project time is 81 days
```

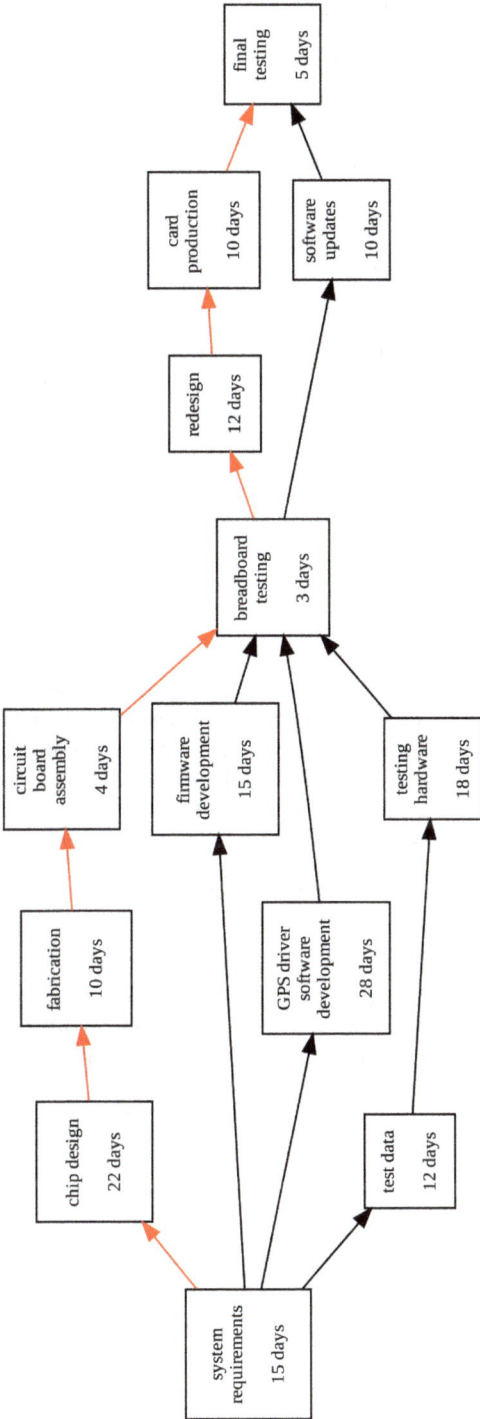

Figure 3.7   Project Network Critical Path

3.5 WIND TURBINE POWER SIMULATION

A power system design requires simulation of wind turbine power with different blade designs across different wind farm locations. The following equations are used for the analysis:

$$A = \pi r^2$$

$$P_{max} = \frac{1}{2} \rho \pi r^2 v^3$$

$$P = P_{max} \cdot C_p$$

where:
A is the swept area
r is the blade radius
ρ is the air density
v is the wind speed
C_p is the power coefficient
P_{max} is the theoretical maximum power using the Betz limit
P is the actual power output.

Listing 3.13 implements the simulation with the function `calculate_power` taking blade parameters and wind speeds. A Rayleigh distribution is used to match collected field data of 10 minute average wind speeds. It performs a day long simulation of intervals and computes total energy generated by the wind turbine.

Listing 3.13   Wind Turbine Power Simulation

```python
import numpy as np

# Constants
PI = 3.14159
AIR_DENSITY = 1.225 # kg/m^3 (standard air density)

def calculate_power(blade_radius, power_coefficient, wind_speed):
    """
    Calculates the power output of a wind blade using the Betz limit ∕
        equation.

    Args:
        blade_radius (float): Radius of the wind blade in meters.
        power_coefficient (float): Power coefficient (Cp) of blade design
        wind_speed (float): Wind speed in meters per second.

    Returns:
        float: Power output of the wind blade in watts.
    """
```

```python
    # Calculate the swept area of the blade
    swept_area = PI * (blade_radius**2)

    # Calculate the theoretical maximum power using the Betz limit (Cp ↲
        = 0.59)
    theoretical_power = 0.5* PI * AIR_DENSITY * wind_speed**3 * ↲
        swept_area

    # Calculate the actual power output
    power_output = theoretical_power * power_coefficient

    return power_output

# Blade parameters
blade_radius = 50 # meters
# Power coefficient (Cp) between 0.3 and 0.45
power_coefficient = 0.4

# Random wind velocities per Rayleigh distribution
# Simulate a day with 10 minute interval averages
total_energy = 0
for interval in range(144):
    wind_speed=np.random.rayleigh(8) # meters per second
    power = calculate_power(blade_radius, power_coefficient, ↲
        wind_speed)
    print(f"Power output at {wind_speed:.1f} m/s is {power/1000:.0f} ↲
        kW")
    total_energy += power/1000 * 10 / 60 # convert to kWh

print(f"Total energy for the day is {total_energy:.0f} kWh")
```

```
Power output at 10.4 m/s is 6828 kW
Power output at 1.5 m/s is 19 kW
Power output at 6.2 m/s is 1430 kW
...
Power output at 14.8 m/s is 19684 kW
Power output at 7.0 m/s is 2049 kW
Power output at 4.5 m/s is 534 kW
Total energy for the day is 287035 kWh
```

3.6   ENGINE PERFORMANCE ANALYSIS

It is desired to know the efficiencies of new automobile engine designs. Listing 3.14 has a generalized class for an automobile engine to vary engine and environmental factors to compute thermal efficiency and power outputs. The thermodynamics analysis covers thermal efficiency, heat added, heat rejected, and the temperature increase due to combustion. Power calculations depend on the thermal efficiency, speed, and torque.

The thermal efficiency analysis uses the following equations:

$$Q_{added} = m_{fuel} \cdot HV, \quad \Delta T = \frac{Q_{added}}{m_{air} \cdot c_p}, \quad Q_{rejected} = m_{air} \cdot \gamma \cdot (T_{final} - T_{initial}),$$

$$\eta_{ideal} = 1 - \frac{1}{r^{\gamma-1}}, \quad \eta_{actual} = \eta_{ideal} \cdot 0.75, \quad m_{air} = V_d \cdot \rho_{air}, \quad m_{fuel} = \frac{m_{air}}{AFR}$$

where:

Q_{added}	Heat added from fuel (J)	m_{air}	Mass of air (kg)
m_{fuel}	Mass of fuel (kg)	c_p	Heat capacity of air (J/kg·K)
HV	Heat value of fuel (J/kg)	$Q_{rejected}$	Heat rejected (J)
ΔT	Temperature increase (K)	γ	Heat ratio of air
T_{final}	Final temperature (K)	$T_{initial}$	Initial temperature (K)
η_{ideal}	Ideal thermal efficiency	η_{actual}	Actual thermal efficiency
r	Compression ratio	V_d	Displacement volume (m³)
ρ_{air}	Air density (kg/m³)	AFR	Air-fuel ratio
		η_v	Volumetric efficiency

The power output is calculated with:

$$P = E_{cycle} \cdot \frac{N}{120}, \quad E_{cycle} = HV \cdot m_{fuel} \cdot \eta_v \cdot \eta_{actual}, \quad \tau = \frac{P \cdot 60}{2\pi N}, \quad P_{hp} = \frac{\tau \cdot N \cdot 2\pi}{60 \cdot 745.7}$$

where:

P	Power output (W)	τ	Torque (Nm)
E_{cycle}	Energy per cycle (J)	N	Engine speed (RPM)
P_{hp}	Power output (hp)		

The Engine class is initialized with attributes for displacement, bore, stroke, and compression ratio to describe the thermodynamic properties of the engine. The temperature and pressure attributes are set to default values. The get_thermal_efficiency method calculates the thermal efficiency of the engine based on the engine's parameters and the fuel heat value, which is used in the get_power_output method to determine the overall engine performance.

The estimate_torque() method estimates the torque based on the thermal efficiency and engine parameters to determine how much of the fuel's energy is converted into useful work. It considers factors for air-fuel ratio and volumetric efficiency in its calculations. The get_power_output() method first calculates an ideal power output based on the input torque and angular velocity, then applies the thermal efficiency to account for real-world engine losses. The class assumes a four-stroke engine cycle in its torque and power estimations, and also includes air temperature and pressure effects during combustion.

Listing 3.14   Engine Thermal Efficiency and Power Analysis

```python
import math

class Engine:
    """
    The Engine class models the thermodynamic properties and ↙
        performance of an internal combustion engine.
    It calculates thermal efficiency, estimates torque, and computes ↙
        power output based on engine parameters.
    Attributes include displacement, bore, stroke, compression ↙
        ratio, and initial temperature and pressure.
    """
    def __init__(self, displacement, bore, stroke, compression_ratio):
        self.displacement = displacement # engine displacement in ↙
            liters
        self.bore = bore # diameter of the cylinder bore in mm
        self.stroke = stroke # length of the piston stroke in mm
        self.compression_ratio = compression_ratio # compression ↙
            ratio of the engine
        self.temperature = 298 # initial temperature of the engine in K
        self.pressure = 101325 # initial pressure of the engine in Pa
        self.thermal_efficiency = None

    def get_thermal_efficiency(self, fuel_heat_value):
        """Calculate the thermal efficiency of the engine"""
        specific_heat_ratio = 1.4# specific heat ratio of air at ↙
            constant pressure and constant volume
        air_mass = self.displacement * 1.225 # mass of air in the ↙
            engine at standard temperature and pressure
        fuel_mass = air_mass / 14.7# stoichiometric air-fuel ratio
        heat_added = fuel_mass * fuel_heat_value # heat added to the ↙
            engine from the combustion of fuel

        # Simulate an increase in temperature due to combustion
        combustion_temperature_increase = heat_added / (air_mass * 100↙
            5) # using specific heat capacity of air at constant ↙
            pressure (J/kgK)
        self.temperature += combustion_temperature_increase

        heat_rejected = air_mass * specific_heat_ratio * ↙
            (self.temperature - 298) # heat rejected from the engine
        self.thermal_efficiency = 1 - (1 / self.compression_ratio ** ↙
            (specific_heat_ratio - 1))

        # Apply a more realistic efficiency factor
        self.thermal_efficiency *= 0.75 # accounting for real-world ↙
            losses

        print(f"Heat added: {heat_added:.2f} J \nHeat rejected: ↙
```

```
                {heat_rejected:.2f} J \nTemperature increase: ↲
                {combustion_temperature_increase:.2f} K")
        return self.thermal_efficiency

    def estimate_torque(self, fuel_heat_value, rpm, ↲
        volumetric_efficiency=0.85):
        """Estimate torque based on engine parameters and thermal ↲
            efficiency"""
        if self.thermal_efficiency is None:
            self.get_thermal_efficiency(fuel_heat_value)

        # Calculate energy per cycle
        cylinder_volume = self.displacement / 1000 # convert to m^3
        fuel_mass_per_cycle = (cylinder_volume * 1.225) / 14.7# ↲
            assuming stoichiometric air-fuel ratio
        energy_per_cycle = fuel_heat_value * fuel_mass_per_cycle * ↲
            volumetric_efficiency * self.thermal_efficiency

        # Estimate torque (assuming 4-stroke engine)
        power = energy_per_cycle * (rpm / 120) # power in watts (2 ↲
            rotations per cycle)
        torque = (power * 60) / (2 * math.pi * rpm) # torque in Nm

        return torque

    def get_power_output(self, torque, rpm):
        """Calculate the power output of the engine"""
        power = (torque * rpm * 2 * math.pi) / 60 # power in watts
        horsepower = power / 745.7# convert watts to horsepower
        return horsepower

# New engine design
new_engine = Engine(3.5, 87, 84, 10.5)
thermal_efficiency = new_engine.get_thermal_efficiency(45e6) # using ↲
    45 MJ/kg as fuel heat value
print(f'Thermal efficiency = {thermal_efficiency:.3f}')

rpm = 6000
estimated_torque = new_engine.estimate_torque(45e6, rpm)
print(f'Estimated torque = {estimated_torque:.1f} Nm')

power_output = new_engine.get_power_output(estimated_torque, rpm)
print(f'Estimated power output = {power_output:.1f} horsepower')
```

```
Heat added: 13125000.00 J
Heat rejected: 18283.58 J
Temperature increase: 3045.99 K
Thermal efficiency = 0.457
Estimated torque = 405.9 Nm
Estimated power output = 342.0 horsepower
```

3.7   UAV QUANTITATIVE FAULT TREE ANALYSIS

The reliability of an Unmanned Aerial Vehicle (UAV) for surveillance missions needs to be analyzed. Quantitative fault tree analysis will be used to quantify the probability of mission data loss. This will provide insight into the most critical components and potential failure points.

The *Systems Engineering Library* (se-lib) has functions for reliability analysis, including both quantitative and qualitative fault trees [19]. It is used in Listing 3.15 to model the potential sources of failure as contributing probabilistic events. A tree structure is specified with nodes for the failure events and edges for their logical relationships. Each node is specified as an *or* gate, an *and* gate, or a basic event with a probability of event occurrence. The *and* and *or* nodes are also specified with their underlying events in the hierarchy. The resulting quantitative fault tree with rolled-up probabilities is shown in Figure 3.8.

In the fault tree model, basic event nodes represent individual component failures. These basic events are then combined using *and* gates and *or* gates to represent the interaction of multiple failures that may cause data loss. An *and* gate signifies that all events beneath it must occur for the fault to happen, while an *or* gate indicates that only one of the events needs to occur. With the specified probabilities of basic events, the overall probability of mission data loss is calculated by propagating the probabilities through the tree structure.

Listing 3.15   UAV Quantitative Fault Tree Analysis

```
import selib as se

# UAV fault tree events and probabilities
uav_fault_tree = [
    ("UAV Mission Data Loss", "or", '', ["Ground Communication ↙
        Loss", "Power Down", "All Sensors Fail"]),
    ('All Sensors Fail', 'and', '', ['RGB Video Fails', 'LiDAR ↙
        Fails', 'Thermal Camera Fails']),
    ('Power Down', 'and', '', ["Main Power Down", "Backup Power ↙
        Down"]),
    ('Ground Communication Loss', 'basic', .003),
    ('Main Power Down', 'basic', .02),
    ('Backup Power Down', 'basic', .08),
    ('RGB Video Fails', 'basic', .001),
    ('LiDAR Fails', 'basic', .001),
    ('Thermal Camera Fails', 'basic', .003),
    ]

# Draw fault tree
se.draw_fault_tree_diagram_quantitative(uav_fault_tree, ↙
    filename="uav_quantitative_fault_tree", format="svg")
```

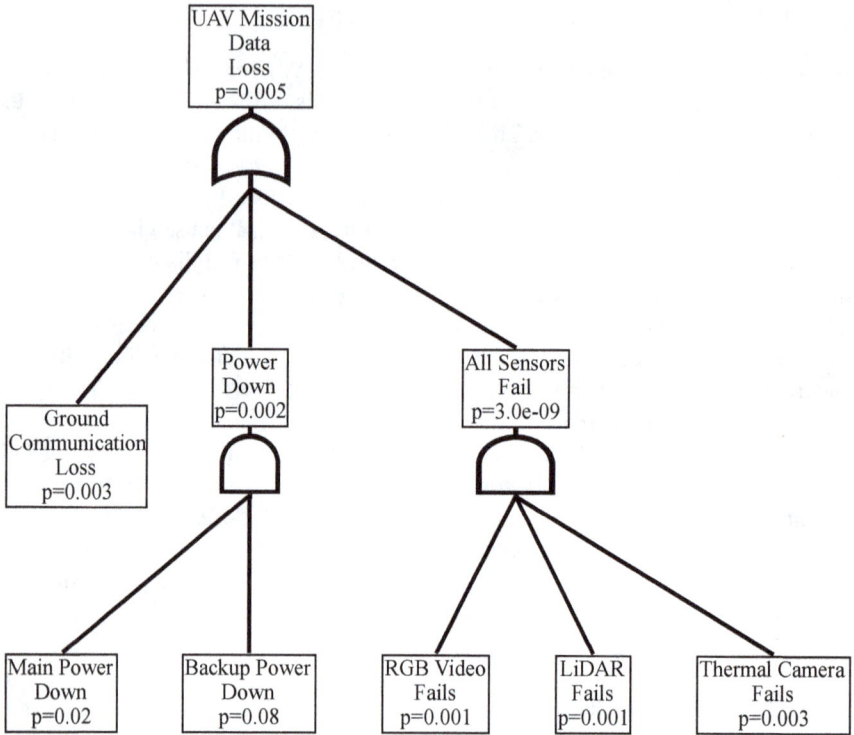

Figure 3.8   UAV Quantitative Fault Tree

3.8   METEOR AND SPACE DEBRIS ENTRY SIMULATION

A country wants to predict the dynamics of a given meteor or space debris entering the atmosphere above it. The same equations for projectile height with air resistance in Section 3.2 will be used for this. The effect of varying air density through the atmosphere is handled with an empirical dataset. A file of air density vs. altitude is read with Pandas that is interpolated from for drag calculations.

     The simulation in Listing 3.16 begins with a meteor entering the edge of earth atmosphere falling vertically. The vertical displacement and velocity state variables are integrated from their derivatives using Euler's method. The effect of ablation is ignored so the meteor mass stays the same.

     The time plots in Figure 3.9 show that as the meteor descends and encounters higher air density at lower altitudes, the drag force increases significantly, decelerating it. The drag force reaches a peak when traveling through dense atmosphere at high velocity. However, as the meteor continues to decelerate, the drag force decreases due to the quadratic dependence of drag on velocity. Even though the air density continues to increase, the reduction in velocity leads to less drag force.

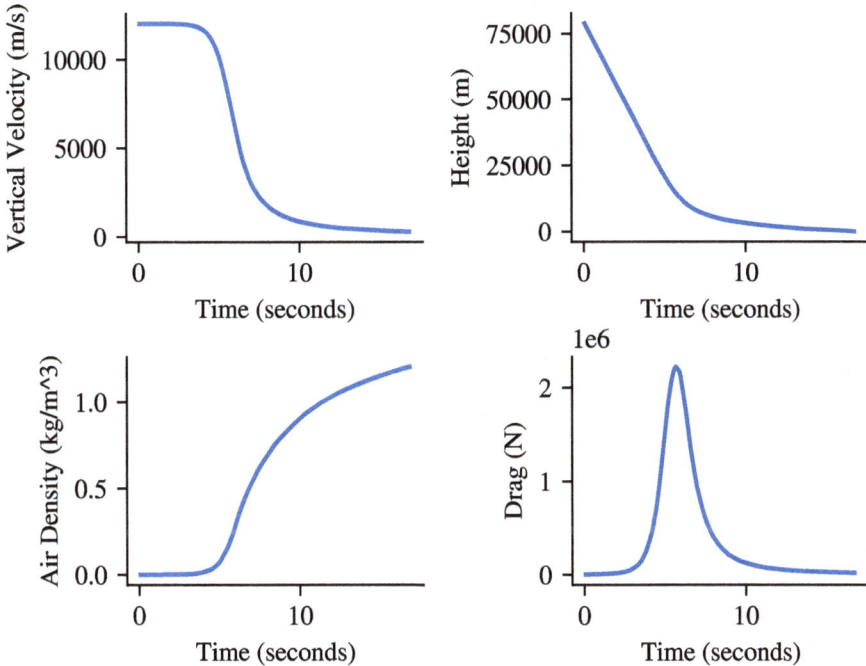

Figure 3.9   Meteor Entry Time Plots

Listing 3.16   Meteor and Space Debris Entry Simulation

```python
import numpy as np
import matplotlib.pyplot as plt
from scipy.interpolate import interp1d
import pandas as pd

# Linear interpolation to get density from altitude
def get_density(altitude, df):
    # use [1:] to ignore first element of dataframe column which is ↙
        its index
    return interp1d(df['Altitude'][1:], df['Density'][1:], ↙
        kind='linear', fill_value="extrapolate")(altitude)

# Dataframe from air density table
df = pd.read_excel('air densities.xlsx')
print(df)

# Constants and initializations
g = 9.81 # m/s^2 (at zero height)
meteor_density = 8000 # kg/m^3
shape_factor = 2
```

```python
mass = 500 # kg
vertical_velocity = 12000 # m/s
acceleration = 0
displacement = 0
height_over_ground = 80000 # m

# Simulation parameters
time = 0
dt = 0.1

# Output lists
air_density_plot = []
height_over_ground_plot = []
vertical_velocity_plot = []
drag_plot = []
time_plot = []

# Time loop
while height_over_ground > 0:
    time += dt
    displacement += vertical_velocity * dt
    vertical_velocity += acceleration * dt

    # Auxiliaries
    gravitation = mass * g # N
    height_over_ground = height_at_start - displacement # m
    air_density = get_density(height_over_ground, df)
    cross_section = (3 * mass / (4 * np.pi * meteor_density)) ** ↙
        (2/3) * np.pi # m^2
    drag = 0.5* cross_section * air_density * shape_factor * ↙
        vertical_velocity ** 2 # N
    total_force = gravitation - drag # N

    # Rates
    acceleration = total_force / mass # m/s^2

    # Store values for plotting
    height_over_ground_plot.append(height_over_ground)
    drag_plot.append(drag)
    vertical_velocity_plot.append(vertical_velocity)
    air_density_plot.append(air_density)
    time_plot.append(time)

# 2 x 2 figure
figure, ((axis1, axis2), (axis3, axis4)) = plt.subplots(2, 2)
axis1.plot(time_plot, vertical_velocity_plot)
axis2.plot(time_plot, height_over_ground_plot)
axis3.plot(time_plot, air_density_plot)
axis4.plot(time_plot, drag_plot)
```

```
axis1.set(xlabel = 'Time (seconds)', ylabel = 'Vertical Velocity ⤸
    (m/s)')
axis2.set(xlabel = 'Time (seconds)', ylabel = 'Height (m)')
axis3.set(xlabel = 'Time (seconds)', ylabel = 'Air Density (kg/m^3)')
axis4.set(xlabel = 'Time (seconds)', ylabel = 'Drag (N)')
plt.subplots_adjust(hspace=0.5, wspace=0.5)
```

3.9 SATELLITE ORBITAL MECHANICS

It is desired to calculate the orbital maneuvers and total fuel cost to transfer a satellite between circular orbits. The Hohmann transfer is a two-impulse maneuver that transfers an orbiting object from one circular orbit to another circular orbit with a different radius. The first impulse increases the object's velocity to raise its apoapsis to the radius of the final orbit. The second impulse decreases the object's velocity to lower its periapsis to the radius of the final orbit.

The *poliastro* library for interactive astrodynamics and orbital mechanics is used to calculate the impulse maneuvers from low earth orbit in Listing 3.17. It computes the delta-v of a transfer maneuver over a range of radii. A delta-v is the change in velocity (delta-v) required for a space maneuver. It is used to determine the amount of propellant required to achieve the delta-v for a propulsive maneuver for a vehicle of given mass and propulsion system. The total cost of fuel can be calculated from the sum of the delta-v impulses.

Orbital parameters are first defined. The `ss_i` variable is an `Orbit` object that represents a circular orbit with attributes. Next output vectors are created to store the delta-v of the first and second impulses for the 1000 radii as the satellite progresses outward. In the main loop over the radii, the call `maneuver ⤸ = Maneuver.hohmann(ss_i, r_f)` returns a `Maneuver` object that represents the Hohmann transfer maneuver from the initial orbit `ss_i` to the final orbit with radius `r_f`. The `impulses` attribute provides a list of tuples which contains the velocity increment (delta-v) and the time of the maneuver. The delta-v impulses are plotted over time in Figure 3.10 normalized to the initial velocity.

Orbits can also be plotted with *poliastro* Matplotlib utilities. The script produces the static orbit trajectories in Figure 3.11 for the Hohmann transfer. The simulation also calculates the time required for each maneuver, illustrating the duration between the impulses. Additionally, perturbative forces like atmospheric drag or non-uniform gravity can be introduced to explore more complex orbital scenarios.

Listing 3.17 Hohmann Transfer Orbital Mechanics

```
from poliastro.bodies import Earth
from poliastro.maneuver import Maneuver
from poliastro.twobody import Orbit
from poliastro.util import norm
from poliastro.plotting.static import StaticOrbitPlotter
from astropy import units as u
```

```python
from matplotlib import pyplot as plt
import numpy as np

# Create a circular orbit around Earth at an altitude of 800 km
orb_i = Orbit.circular(Earth, alt=800 * u.km)

# Entire transfer cost and time to 80000 km
hoh = Maneuver.hohmann(orb_i, 80000 << u.km)
print('Total Cost =', hoh.get_total_cost())
print('Total Time =', hoh.get_total_time())

 # Retrieve the maneuver transfer and resulting orbits
orb_a, orb_f = orb_i.apply_maneuver(hoh, intermediate=True)

# Initial radius and velocity of the orbit
r_i = orb_i.a.to(u.km)
v_i_vec = orb_i.v.to(u.km / u.s)
v_i = norm(v_i_vec)

# Create a vector of final radii from 1 to 100 times the initial radius
N = 100 # The number of final radii to calculate
dv_a_vector = np.zeros(N) * u.km / u.s # A vector to store the ↙
    delta-v of the first impulse
dv_b_vector = dv_a_vector.copy() # A vector to store the delta-v of ↙
    the second impulse
r_f_vector = r_i * np.linspace(1, 100, num=N) # A vector of final ↙
    radii from 1 to 100 times the initial radius

# Iterate over the final radii to calculate delta-vs and magnitudes
for ii, r_f in enumerate(r_f_vector):
    # Calculate the Hohmann transfer maneuver from the initial ↙
        circular orbit to orbit with radius r_f
    maneuver = Maneuver.hohmann(orb_i, r_f)

    # Retrieve the impulses (delta-v) for the first and second ↙
        maneuvers from the Hohmann transfer maneuver
    # The impulse is a tuple containing the velocity increment ↙
        (delta-v) and the time of the maneuver
    (_, dv_a), (_, dv_b) = maneuver.impulses

    # Calculate the norm (magnitude) of the delta-v vector for the ↙
        first and second impulses
    dv_a_vector[ii] = norm(dv_a)
    dv_b_vector[ii] = norm(dv_b)

# Plot the delta-v of the impulses and the total
fig, ax = plt.subplots()
ax.plot((r_f_vector), (dv_a_vector / v_i), "blue", label="First ↙
    impulse")
```

```
ax.plot((r_f_vector), (dv_b_vector / v_i), "red", label="Second ↙
    impulse")
ax.plot((r_f_vector), ((dv_a_vector + dv_b_vector) / v_i), "green", ↙
    label="Total $\Delta v$",)

# Set axis limits and labels
ax.set(ylim = (0, 0.8), xlabel ="$R$ (km)", ylabel= "$\Delta v_a / ↙
    v_i$")
ax.legend()

# Create static plot of orbits using Matplotlib backend
plotter = StaticOrbitPlotter()
plotter.plot(orb_i, label="Initial orbit")
plotter.plot(orb_a, label="Transfer orbit")
plotter.plot(orb_f, label="Final orbit")
plt.legend(title="Orbits and Epochs", loc='center', ↙
    bbox_to_anchor=(0.5, 0.75))
```

```
Total Cost = 3.9698030853042168 km / s
Total Time = 45284.269274637365 s
```

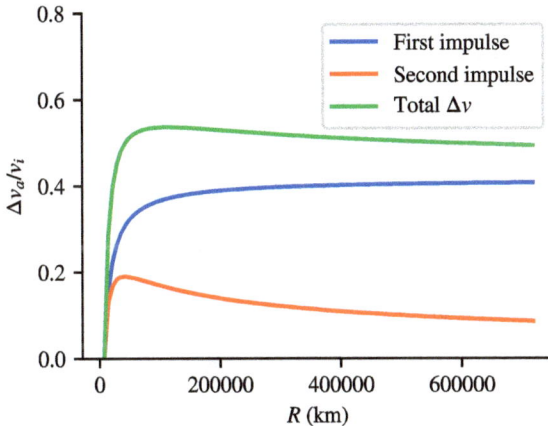

Figure 3.10   Hohmann Transfer Delta-vs

3.10   FLUID COMPRESSIBILITY FACTOR ANALYSIS

Gas compressibility factors are used to design industrial process equipment for fluid flow. The critical compressibility factor z_c is a dimensionless quantity that describes the deviation of a real gas from ideal gas behavior at its critical point. It is used to design pipes and valves for different operating conditions in the plant. In this

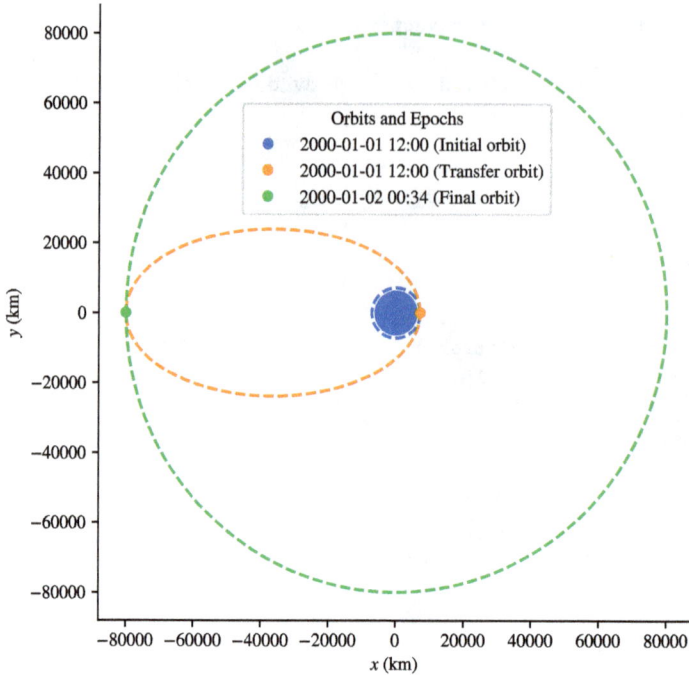

Figure 3.11   Hohmann Transfer Orbits

example, heptane will be used for cleaning circuit cards and other electronics in the FastCircuit plant.

The equation for z_c is:

$$z_c = \frac{P_c \mu}{R \rho_c T_c}$$

where:

P_c is the critical pressure of the gas

μ is the mean molecular weight of a gas particle in units of the mass of a hydrogen atom

R is the ideal gas constant

ρ_c is the critical density of the gas

T_c is the critical temperature of the gas.

Cantera is a library for chemical kinetics, thermodynamics, and transport processes used in this example. In the following, z_c is calculated and then used to plot the contour lines for the compressibility factor isolines at different temperature and pressure values. The output graph in Figure 3.12 shows the compressibility for the desired range of temperatures and pressures.

Listing 3.18   Fluid Compressibility Factor Analysis

```python
import cantera as ct
import numpy as np
import matplotlib.pyplot as plt

# Define fluid
fluid = ct.Heptane()

# Define range of pressures
p_min = 0.05e6 # Pa
p_max = 10e6  # Pa
num_p = 20    # Number of pressure values to consider
pressures = np.linspace(p_min, p_max, num_p)

# Define temperature range
T_min = 300
T_max = 1000
num_T = 20
T = np.linspace(T_min, T_max, num_T)

tc = fluid.critical_temperature
pc = fluid.critical_pressure
rc = fluid.critical_density
mw = fluid.mean_molecular_weight
zc = pc * mw / (rc * ct.gas_constant * tc) # critical ↙
    compressibility factor

z = np.zeros((num_T, num_p))
for i in range(num_T):
    for j in range(num_p):
        T_curr = T[i]
        p_curr = pressures[j]
        fluid.TP = T_curr, p_curr
        rho = fluid.density
        z[i, j] = p_curr * mw / (rho * ct.gas_constant * T_curr)

# Plot compressibility factor isolines
# Get the min and max values
x_min, x_max = np.min(T), np.max(T)
y_min, y_max = np.min(pressures / 1e6), np.max(pressures / 1e6)

# Scale the x and y axes to the actual range of data
plt.xlim(x_min, x_max)
plt.ylim(y_min, y_max)

# Create contour lines
cs = plt.contour(T, pressures / 1e6, z, levels=[zc - 0.1 , zc, zc + 0↙
    .1 ])
```

```
# Create labels for the contour lines
labels = [f'z={zc-0.1:.2f}', f'z={zc:.2f}', f'z={zc+0.1:.2f}']

# Add the legend with specified labels
plt.legend(cs.legend_elements()[0], labels)

plt.xlabel('Temperature (K)')
plt.ylabel('Pressure (MPa)')
```

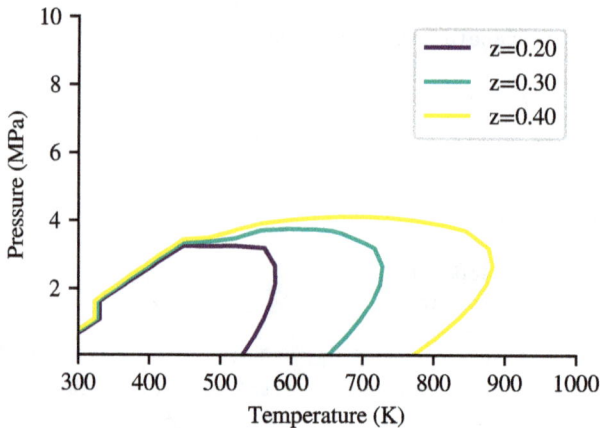

Figure 3.12   Heptane Compressibility Isolines

3.11   MACHINE LEARNING

Python has several powerful machine learning libraries for a wide range of engineering applications. Scikit-Learn [17] is widely-used with extensive tools for supervised and unsupervised learning tasks such as classification, regression, clustering, and dimensionality reduction. TensorFlow is oriented for deep learning with an ecosystem of tools and libraries often used for image recognition, natural language processing, and speech recognition. PyTorch [16] is also used for deep learning known for its dynamic computation graph and a wide range of pre-trained models that can be easily used for transfer learning. Keras is a high-level neural network API that can run on top of TensorFlow and others. It provides an easy-to-use interface for building and training deep learning models that can be fine-tuned for specific tasks.

Regression and classification for prediction purposes are two common types of machine learning algorithms demonstrated here. Regression is used to predict continuous values. It minimizes the difference between the predicted values and the actual values, by finding the line or curve that best fits the data. Classification is used to predict categorical values. It aims to separate data into different classes based on the features, by finding the decision boundary that separates the classes.

3.11.1 REGRESSION

This example uses Pandas to load a dataset from a CSV file. The target variable is separated from the feature variables, and the data is split into training and testing sets using the train_test_split function. A linear regression model is then trained on the training data using the fit method of the LinearRegression class. Finally, the model is evaluated on the test data by calculating the mean squared error (MSE) between the predicted values and the actual values.

Listing 3.19 Machine Learning Linear Regression

```python
import numpy as np
import pandas as pd
from sklearn.model_selection import train_test_split
from sklearn.linear_model import LinearRegression

# Load the engineering dataset from a CSV file
df = pd.read_csv("engineering_data.csv")
X = df.drop('Target_Variable', axis=1).values
y = df['Target_Variable'].values

# Split the data into training and testing sets
X_train, X_test, y_train, y_test = train_test_split(X, y, ⤸
    test_size=0.5)

# Train a linear regression model
reg = LinearRegression().fit(X_train, y_train)

# Evaluate the model on the test data
y_pred = reg.predict(X_test)
mse = np.mean((y_pred - y_test)**2)

# Print the regression coefficients and intercept
print(f"Regression Coefficients: {reg.coef_}")
print(f"Regression Intercept: {reg.intercept_}")

# Print the mean squared error and R-squared value
r_squared = reg.score(X_test, y_test)
print(f"Mean Squared Error: {mse}")
print(f"R-Squared: {r_squared}")
```
```
Regression Coefficients: [1.01796466 1.01018375]
Regression Intercept: 2.592130022284877
Mean Squared Error: 3.2453278780191535
R-Squared: 0.9747964656724574
```

3.11.2   MACHINE LEARNING FOR ELECTRONICS CLASSIFICATION

FastCircuit desires to automate the identification of electronic circuit cards moving on conveyors to route for further inspection, maintenance, and refurbishing. Machine learning will be used to shorten overall processing and routing times. The example in Listing 3.20 uses Scikit-Learn for the automated classification of three versions of circuit cards from laser measurements.

Each card version has potential differences due to temporal manufacturing processes, and local parts and chip customizations leading to variations. Furthermore used cards may have occluding deposition and dirt. They are placed on a rapidly moving conveyor with laser measurements taken of the thickness, length, heat gap width, and an overall reflectance value. Imperfect measurement variations are due to the conveyer rapidity, and uneven cards laid due to geometry and jutting chips.

It is necessary to first have actual measurements and human identification of the correct card types A, B, or C for training the algorithm. *Pandas* is used to read the data in "boards.csv" and populate a dataframe to be sent to *sklearn*. The *sklearn* library is used for the decision tree classifier splitting it into a training dataset and testing set, the proportions of which can be varied.

By using 90% of the data for training and 10% for prediction, the results shown are very good with the data. The output confusion matrix shows that the classification model has made mostly correct predictions on the test data, as indicated by the large number of samples along the diagonal of the matrix. The matrix is a 3x3 matrix since there are 3 classes in the dataset.

Diagonal elements in the confusion matrix represent the number of samples that were correctly classified for each class. E.g., the cell (1,1) indicates that 44 samples that belonged to class A were correctly classified and hence its precision is 1.0. The off-diagonal elements represent the misclassified; thus the cell (2,3) indicates that 3 samples that belonged to the second class were misclassified as belonging to the third class.

The *seaborn* library pairplot function is also demonstrated and shown in Figure 3.13. It isn't required for the machine learning algorithm but is convenient for visualizing multidimensional data and their relationships in many applications. The pairplot function creates a grid of plots that show the pairwise relationships between the variables in a dataset. The diagonal plots in the grid show the univariate distribution of each variable. The off-diagonal plots show the bivariate relationships between each pair of variables shown as scatter plots here. The three types of cards are color coded to help visualize their spatial differences.

Listing 3.20   Machine Learning for Circuit Board Classification

```
import numpy as np
import pandas as pd
from sklearn.model_selection import train_test_split
from sklearn.tree import DecisionTreeClassifier
from sklearn.metrics import classification_report, confusion_matrix, ↙
    accuracy_score
import matplotlib.pyplot as plt
```

```python
import seaborn as sns

# Read the circuit card dataset from a CSV file
boards = pd.read_csv("boards.csv")
X = boards.iloc[:, :-1].values
y = boards.iloc[:, -1].values

# Split the data into training and testing sets
X_train, X_test, y_train, y_test = train_test_split(X, y, ↲
    test_size=0.9)

# Train a decision tree classifier
clf = DecisionTreeClassifier()
clf.fit(X_train, y_train)

# Make predictions on the test data
y_pred = clf.predict(X_test)

# Print classification report and confusion matrix
print("Classification Report:")
print(classification_report(y_test, y_pred))
print("Confusion Matrix:")
print(confusion_matrix(y_test, y_pred))

# Calculate and print accuracy score
accuracy = accuracy_score(y_test, y_pred)
print(f"Accuracy: {accuracy}")

# create pairwise plot with seaborn
sns.pairplot(boards, hue='version')
```

```
Classification Report:
              precision    recall  f1-score   support

           A       1.00      1.00      1.00        44
           B       0.93      0.90      0.91        48
           C       0.89      0.93      0.91        43

    accuracy                           0.94       135
   macro avg       0.94      0.94      0.94       135
weighted avg       0.94      0.94      0.94       135

Confusion Matrix:
[[44  0  0]
 [ 0 43  5]
 [ 0  3 40]]
Accuracy: 0.9407407407407408
```

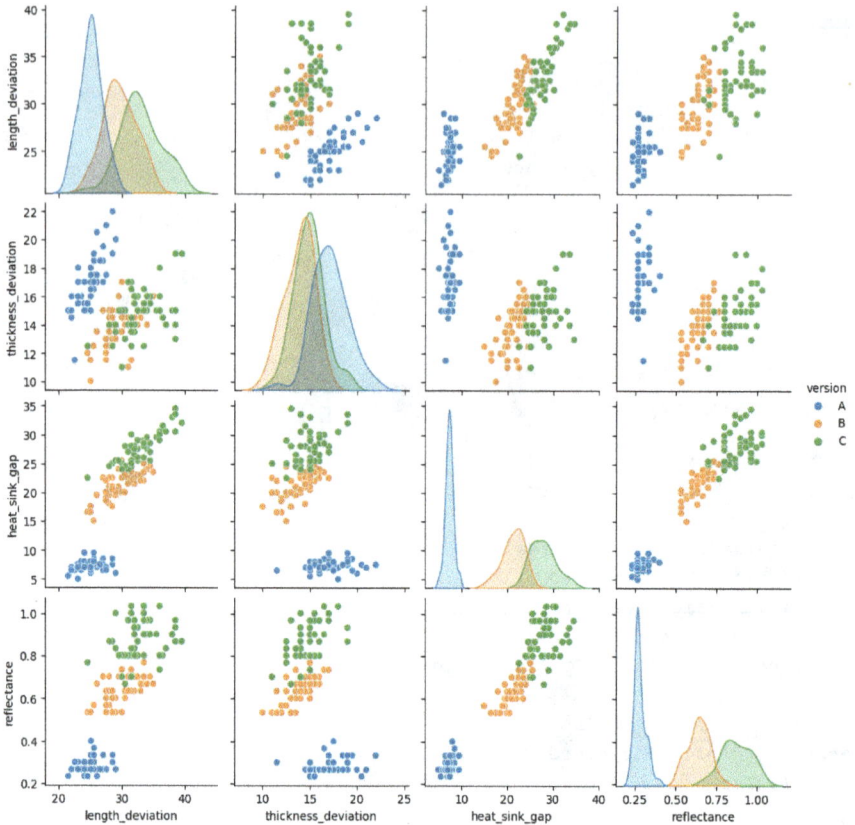

Figure 3.13   Training Set Pairplots for Electronic Boards

3.12   ROBOT PATH CONTROL

The 2-dimensional path of a manufacturing transportation robot is simulated to test the kinematic input commands prior to controlling actual robots. It is programmed to leave a docking area, proceed to assembly stations, perform a figure eight path around the stations to transfer items, and return. The simulation is used to test the control inputs safely before uploading into the robot hardware.

The equations used for forward kinematics of a robot are:

$$\dot{x} = v\cos(\theta)$$
$$\dot{y} = v\sin(\theta)$$
$$\dot{\theta} = w$$
$$x(t) = x(t-1) + \dot{x}dt$$
$$y(t) = y(t-1) + \dot{y}dt$$
$$\theta(t) = \theta(t-1) + \dot{\theta}dt$$

where:

x is x position of the robot

y is y position of the robot

$\dot{x}$ is x velocity of the robot

$\dot{y}$ is y velocity of the robot

θ is heading angle of the robot

v is linear velocity of the robot

w is angular velocity of the robot

dt is time step.

The simulate_robot() function calls the `forward_kinematics` function in a loop to update the robot state over time per the equations. It returns the x, y, and theta values for each time step given the control inputs.

Listing 3.21   Robot Path Control

```python
import numpy as np
import matplotlib.pyplot as plt

def forward_kinematics(x, y, theta, v, w, dt):
    # Update robot state using forward kinematics
    x_dot = v * np.cos(theta)
    y_dot = v * np.sin(theta)
    theta_dot = w * np.pi / 180
    x += x_dot * dt
    y += y_dot * dt
    theta += theta_dot * dt
    return x, y, theta

def simulate_robot(initial_x, initial_y, initial_theta, ↙
    control_inputs, dt):
    # Simulate robot motion over time
    x, y, theta = initial_x, initial_y, initial_theta
    path = [(x, y, theta)]
    t = 0
    for u in control_inputs:
        x, y, theta = forward_kinematics(x, y, theta, *u, dt)
        # print(*u) # uncomment to print state variables over time
        path.append((x, y, theta))
        t += dt
    return path

# Initialize robot state
initial_x = 0
initial_y = 0
initial_theta = np.radians(0)

# Define control inputs as a list of tuples with linear velocity and ↙
```

```
                angular velocity for each time step.

# Control inputs to proceed to stations, perform a figure 8 and ↵
    return on same path.
# List multipliers are the time durations for the control inputs
controls = [(5, 0)] * 5 + [(5, 180)] * 5 + [(5, 0)] * 5 + [(1, 90)] ↵
    * 100 + [(1, -90)] * 100

# Simulate robot motion over time
dt = .1
trajectory = simulate_robot(initial_x, initial_y, initial_theta, ↵
    controls, dt)

# Print final robot state
final_x, final_y, final_theta = trajectory[-1]
print("Final robot state:")
print(f"x: {final_x:.2f}, y: {final_y:.2f}, theta: {final_theta:.2f}")

# Plot robot trajectory
x_vals, y_vals, theta_vals = zip(*trajectory)
plt.plot(x_vals, y_vals, '-o')
plt.xlabel("X position")
plt.ylabel("Y position")
plt.title("Robot Path")
```

3.13   MANUFACTURING DISCRETE EVENT SIMULATION

FastCircuit wants to reduce the testing times of circuit boards. A simple discrete event model is written that accounts for probabilistic waiting times and delays. The number of test stations is varied to help decide if more stations should be invested in.

The *se-lib* library also provides functions for systems modeling and simulation [19]. It is used in Listing 3.22 for a discrete event network model of the testing process.

Listing 3.22   Manufacturing Test Discrete Event Simulation

```
# Discrete event model of board testing

import selib as se
import numpy as np
import matplotlib.pyplot as plt

def board_testing_model(num_testers):
    """
    A discrete event simulation model for testing electronic boards ↵
        in a manufacturing process.
```

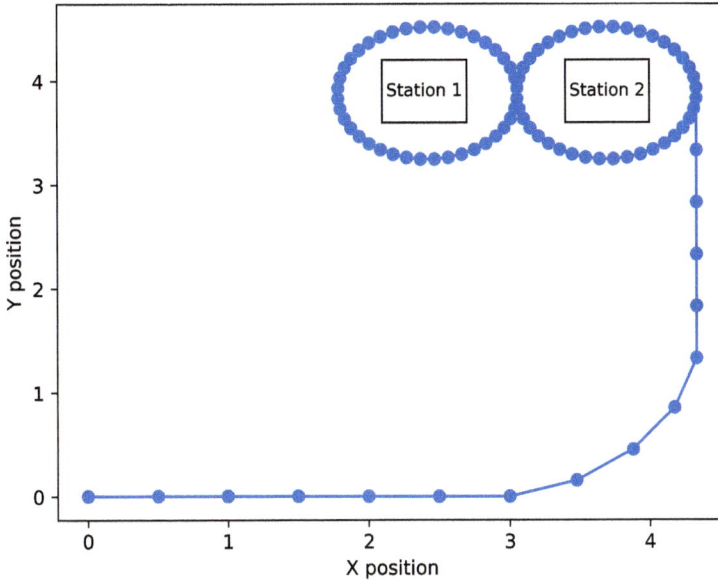

Figure 3.14   Robot Path

```
The model simulates the arrival of electronic boards, their ↵
    processing through
stations, and the completion of the testing process. The number ↵
    of available testers
(servers) can be specified to vary the testing capacity.

Args:
    num_testers (int): The number of testers (servers) available ↵
        for testing the boards.

Returns:
    tuple:
        - model_data (dict): A dictionary containing the overall ↵
            model data, including
          timing and resource usage statistics.
        - entity_data (list of dict): A list of dictionaries, ↵
            each containing detailed
          data for each entity (board) that was processed in the ↵
            simulation.
    """
se.init_de_model()
```

```python
    se.add_source('incoming boards',
            entity_name="board",
            num_entities = 100,
            connections={'firmware installation': 1},
            interarrival_time='np.random.exponential(.6)')

    se.add_delay('firmware installation',
            connections={'testing station': 1},
            delay_time='np.random.uniform(.5, 1)')

    se.add_server(name='testing station',
            connections={'passed boards': .95, 'failed boards': .05},
            service_time='np.random.uniform(.1, .5)',
            capacity = num_testers)

    se.add_terminate('passed boards')
    se.add_terminate('failed boards')
    model_data, entity_data = se.run_model(verbose=False)
    return model_data, entity_data

# Simulation parameters
num_runs = 1000
testers = [1, 2, 3]

# Monte Carlo simulation and collection of average waiting times
average_waiting_times = [] # list of lists for each testing scenario

for num_testers in testers:
    scenario_average_waiting_times = []

    for run in range(num_runs):
        model_data, board_data = board_testing_model(num_testers)
        average_waiting_time = np.mean(model_data['testing ↲
            station']['waiting_times'])
        scenario_average_waiting_times.append(average_waiting_time)
    average_waiting_times.append(scenario_average_waiting_times)

# Create box plots for different numbers of test stations
fig, ax = plt.subplots()
ax.boxplot(average_waiting_times)
ax.set(xlabel = "Number of Test Stations", ylabel="Average Waiting ↲
    Time (Hours)", title = "Average Waiting Times for Different ↲
    Numbers of Testers", xticklabels=testers)
```

The structure of the model is visualized in Figure 3.15 showing the network paths of the boards entering and leaving the testing process. It can be drawn using the `draw_model_diagram()` function after the model elements are specified.

```python
se.draw_model_diagram()
```

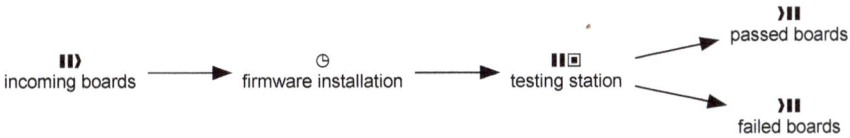

Figure 3.15   Manufacturing Discrete Event Model Diagram

Incoming boards are generated per an exponential inter-arrival time. The testing station is modeled as a server with a uniformly distributed service time and variable capacity for the number of testers. Queues form at the test stations where the waiting times are collected from. The program performs a Monte Carlo analysis simulating 100 boards across 1000 runs. The averages for each run are appended to lists for each testing scenario.

This example performs many runs to gather Monte Carlo statistics. The per-run statistics are computed from many individual events which are not desirable to view in Monte Carlo mode, but run data can be instrumented to see what is transpiring. Detailed event output for a single run can be shown with a verbose flag as below, while other model data and event timing details for each entity are also available in output dictionaries.

```
se.run_model(verbose=True)
0.3212: board 1 entered from incoming boards
0.3212: board 1 incoming boards -> firmware installation
0.3451: board 2 entered from incoming boards
...
2.8636: board 3 leaving system at passed boards
2.8636: board 7 granted testing station resource waiting time 1.1350
2.8701: board 13 entered from incoming boards
...
```

The Monte Carlo simulation results are displayed in Figure 3.16. For each scenario, it shows a boxplot of 1000 samples to summarize the distribution of waiting times.

3.14   ROBOT COMPETITION SIMULATION

A team of engineering students wants to assess strategies for a robot competition. The battle involves rounds where projectiles are fired simultaneously against competitors until the last robot remains. It assumes that each robot will fire at one other random remaining robot during a round. Each robot has attributes for its health and attack power against opponents. The health of a robot degrades with each hit until failure per the following rate of change:

$$\frac{dH_i(t)}{dt} = -\sum_{j=1}^{N} P_{ij}(t)A_jH_j(t)$$

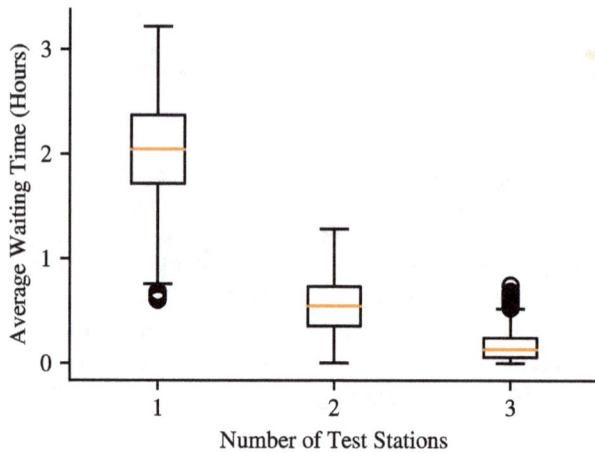

Figure 3.16   Waiting Times Boxplots vs. Test Stations

where:

$H_i(t)$ is the health of robot i at time t

N is the total number of robots

$P_{ij}(t)$ is the probability that robot j chooses robot i as its target at time t

A_j is the attack power of robot j

$H_j(t)$ is the health of robot j at time t.

The simulation model in Listing 3.23 has two major classes for the simulation. It enables the creation of multiple robots with health and attack power attributes with the Robot class. The Battle class allows for inclusion of any number of robots and runs a simulated battle. Outputs are shown for a simple battle with three robots. A plot of the robot health over salvos is in Figure 3.17.

Listing 3.23   Robot Battle Simulation

```
import random
import matplotlib.pyplot as plt


class Robot:
    """
    A robot class with health, attack, and fire_weapon methods.
    """

    def __init__(self, name, health, attack):
        """
        Initializes a robot with the given name, health, and attack ⤸
            power.
        """
```

```python
        self.name = name
        self.health = health
        self.max_health = health
        self.attack = attack

    def fire_weapon(self, target):
        """
        Fires a weapon at the target robot, reducing its health by ↙
            the attack power.
        """
        damage = random.randint(0, self.attack)
        target.health -= damage
        target.health = max(0, target.health) # Ensure health doesn't ↙
            go below 0
        print(f"{self.name} fires at {target.name} and causes ↙
            {damage} damage")
        if target.health <= 0:
            print(f"{target.name} has been destroyed!")

class Battle:
    """
    A class to manage the robot battle.
    """
    def __init__(self):
        self.robots = []
        self.health_history = []

    def add_robot(self, robot):
        """
        Adds a robot to the battle.
        """
        self.robots.append(robot)

    def attack(self):
        """
        Simulates a round attack by having each robot fire at a ↙
            random target.
        """
        for attacker in self.robots:
            if attacker.health <= 0:
                continue
            possible_targets = [robot for robot in self.robots if ↙
                robot != attacker and robot.health > 0]
            if possible_targets:
                target = random.choice(possible_targets)
                attacker.fire_weapon(target)

    def record_health(self):
        """
```

```
        Records the current health of all robots.
        """
        self.health_history.append([robot.health for robot in ↙
            self.robots])

    def plot_health_chart(self):
        """
        Plots a step chart of robot health over rounds
        """
        rounds = range(1, len(self.health_history) + 1)
        robot_names = [robot.name for robot in self.robots]

        fig, ax = plt.subplots()

        for i, robot in enumerate(self.robots):
            health_data = [round[i] for round in self.health_history]
            ax.step(rounds, health_data, label=robot.name)

        ax.set_xlabel('Round')
        ax.set_ylabel('Robot Health')
        ax.set_xticks(rounds)
        ax.legend()

        plt.ylim(0, max(robot.max_health for robot in self.robots))
        plt.tight_layout()
        plt.savefig("robot_health_chart.pdf")
        plt.show()

    def run(self):
        """
        Runs the battle until only one robot remains.
        """
        round_num = 0
        while sum(1 for robot in self.robots if robot.health > 0) > 1:
            round_num += 1
            print(f"\nRound {round_num}:")
            self.attack()
            self.record_health()

        winner = next(robot for robot in self.robots if robot.health ↙
            > 0)
        print(f"\n{winner.name} wins the battle!")

        self.plot_health_chart()

# Create a battle and add robots
battle = Battle()
battle.add_robot(Robot("Trojan", 100, 20))
battle.add_robot(Robot("Bruin", 100, 20))
```

```
battle.add_robot(Robot("Fighting Irishman", 100, 20))

# Run the battle
battle.run()
Round 1:
Trojan fires at Bruin and causes 11 damage
Bruin fires at Fighting Irishman and causes 15 damage
Fighting Irishman fires at Trojan and causes 2 damage

Round 2:
Trojan fires at Bruin and causes 0 damage
Bruin fires at Trojan and causes 8 damage
Fighting Irishman fires at Bruin and causes 7 damage

Round 3:
Trojan fires at Fighting Irishman and causes 10 damage
Bruin fires at Trojan and causes 7 damage
Fighting Irishman fires at Bruin and causes 12 damage
...
Round 12:
Trojan fires at Fighting Irishman and causes 9 damage
Fighting Irishman has been destroyed!

Trojan wins the battle!
```

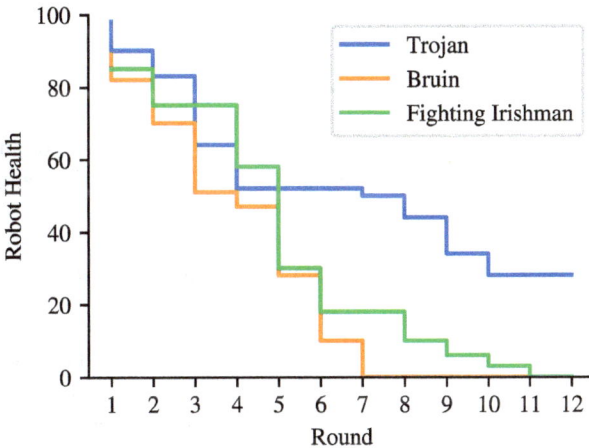

Figure 3.17   Robot Health Over Rounds

3.15   EXERCISES

Modify the projectile with air resistance example for any of the following.

1. Vary the drag coefficients associated with different projectile shapes. Visualize the trade-offs between range, height, and overall motion of the projectile for different coefficients.

2. Use numerical optimization in SciPy or another method to find the optimal launch angle that maximizes the range of the projectile, accounting for air resistance.

Modify the economic present value routines for any of the following:

3. Allow for compounding at the beginning of periods and flexible period lengths shorter than one month.

4. Use more flexible dictionary inputs for the cash flows. Use time periods as the primary keys, and include sub-dictionaries for the inflows and outflows during each time period. Allow for multiple streams of cash flows.

5. Refine or replace the engine performance model. Some further considerations include:

   • Adding factors for mechanical efficiency to account for frictional losses.
   • Incorporating more detailed combustion and heat loss models.
   • Including temperature-dependent properties for fluids.
   • Accounting for varying air-fuel ratios and ignition timing.

6. Extend the program to create a 3D plot of the compressibility factor as a function of both temperature and pressure for a chosen fluid. Instead of contour lines, the compressibility factor should be represented as a continuous surface, where the height of the surface (the z-axis) shows how Z changes across the entire range of temperatures and pressures. Analyze how the compressibility factor changes across a range of pressures and temperatures. Identify regions where the fluid behaves as an ideal gas and where it deviates significantly.

7. Augment the critical path analysis example to determine late start times and slack times for a given set of activities. Answer the following questions: 1) What is the latest time each activity can start without delaying the overall project? 2) How much slack time is available for each activity?

   The slack for an activity can be used to determine its early start (ES), early finish (EF), late start (LS), and late finish (LF) times. Enhance the task network visualization to add fields in the task boxes for ES, EF, LS, LF, and slack. Fine-tune the aesthetics and styling.

8. Modify the robot battle simulation for any of the following:

- Revise the simulation to include salvo attacks, where multiple projectiles are fired in each round. Account for the degradation of capabilities affecting hit probabilities and lethality for each projectile fired.

- Create teams of robots with different attributes, and vary their attack power and health. Simulate many battles, and analyze which strategy (high attack vs. high health) performs better on average.

- Extend the simulation to implement strategic targeting algorithms. For example, a robot could target the robot with the lowest health or the highest attack power. Also vary the starting health and attack power of each robot. Run Monte Carlo simulations with each targeting strategy, and compare the probabilistic results to determine which strategy leads to better outcomes in terms of survivability or winning probability. Visualize the results using boxplots, histograms, or other statistical tools to compare the effectiveness of different robot designs or strategies.

9. Conduct linear regression and hypothesis testing, and derive confidence intervals to analyze engineering data. Collect a dataset related to an engineering problem, and perform linear regression on it. After fitting the model, conduct a hypothesis test to determine whether the slope of the regression line is significantly different from zero. Calculate the 95% confidence intervals for both the slope and intercept. Graph the results, and draw conclusions about the engineering phenomenon represented by the data. Optionally, visualize the p-values using Z-distribution plots.

10. Extend the meteor entry program for the following:

- Allow objects to enter at an angle to the atmosphere, and account for Earth's curvature.

- Model ablation effects where the object's mass decreases over time.

- Compare the effects of different shape factors, mass, and mass densities.

11. Enhance the existing Hohmann transfer program to optimize the transfer based on minimal fuel consumption. Use numerical optimization to adjust the semi-major axis of the transfer orbit and find the most efficient trajectory in terms of delta-v. The current example assumes fixed orbital radii and provides a two-impulse transfer solution, calculating the required delta-v and transfer time without optimization. Adjusting the semi-major axis may result in a more fuel-efficient transfer.

12. Extend the existing machine learning regression example by testing other regression models from Scikit-Learn or another machine learning library. Assess their prediction performance and goodness-of-fit. Compare each model's performance to the original linear regression using metrics for Mean Squared

Error (MSE), Adjusted R-squared, Standard Error, and p-values for each co-efficient. Evaluate how well the models fit the data, and analyze the statistical significance of each regression coefficient. Perform this analysis on the provided dataset or another dataset of your choice.

13. Test different machine learning algorithms from Scikit-Learn against the existing classification model for electronic boards, and evaluate the results. Compare their performances using the metrics precision, recall, and F1-score.

3.16   ADVANCED EXERCISES

1. Integrate fault tree analysis with Monte Carlo simulation and perform a probabilistic analysis for a given system. Use probability distributions to model basic event probabilities. These can be simple uniform, triangular, or PERT distributions. Use Monte Carlo analysis to run the model for at least 1,000 iterations to develop distributions for the AND gates, the OR gates, and the top-level fault probability. Plot the resulting distributions.

2. Add additional gate types to the *se-lib* fault tree functions.

3. Develop the capability to compute and visualize reliability block diagrams as an alternative to fault trees.

4. Optimize the design of a wind turbine by extending the example to vary the power coefficient, location (wind speed), and blade radius to maximize energy output over a year. Use generated or actual wind speed data for multiple locations to simulate realistic wind conditions. Visualize the results as 3D surface plots for each location, showing energy output on the z-axis against power co-efficient and blade radius on the x- and y-axes. Note that structural limits are not modeled. Vary the blade radius in a feasible range from 30m to 70m and power coefficient from 0.3 to 0.5.

5. Add a cost-benefit analysis to the extended wind turbine exercise from above. Incorporate estimated costs for increasing the blade radius, building turbines in high-wind locations, and optimizing the turbine design.

   a. Make the critical path analysis stochastic by using probability distributions for task durations. Allow for multiple distribution types, including uniform, triangular, normal, lognormal, and others.
   Conduct a Monte Carlo analysis and gather statistics on individual task durations, slack times, and total project duration. Generate frequency histograms for the output variables and a cumulative distribution for the total project duration.

   b. Create a master class for the dynamics of orbiting objects, with derived classes covering human-engineered satellites (as in the Hohmann transfer program) and meteors in the Earth's vicinity.

c. Create a master class for the dynamics of orbiting objects, with derived classes for controlled objects like human-engineered satellites (as in the Hohmann transfer program) and natural objects like meteors. For satellites, account for the ability to change orbits using delta-v maneuvers. For both meteors and defunct satellites, model their uncontrolled descent through the atmosphere.

d. Extend the Hohmann transfer program to include a bi-elliptic transfer option to assess fuel and time trade-offs. Implement the bi-elliptic transfer for the same initial and final orbits, comparing the total delta-v, time required for each transfer, the number of impulses, and the positions of the burns. Vary the ratio of the final orbit's radius to the initial orbit's radius over a wide range (e.g., LEO to GEO), and determine when the bi-elliptic transfer is more efficient than the Hohmann transfer. Visualize the trade-offs by plotting delta-v versus transfer time for varying final orbit radii. Include annotations to highlight when the bi-elliptic transfer becomes more efficient than the Hohmann transfer.

e. Apply the machine learning classification task to another engineering application, such as classifying different types of objects based on physical measurements. Use the machine learning process of data preprocessing, training, and testing to implement a classification. Experiment with different training and testing data splits (e.g., 70/30, 80/20). Visualize the results using a confusion matrix and performance metrics.

f. Extend the robot path control to simulate obstacle avoidance. Introduce obstacles along the robot's path, and modify the control inputs dynamically to avoid collisions. The robot should adjust its angular velocity to navigate around obstacles without deviating from its overall objective.

g. Develop an optimization algorithm for controlling the robot's path to minimize energy consumption or travel time. Use it to find the most efficient route between two points in the workspace, taking obstacles into account.

h. Create a real-time control interface that allows users to adjust the robot's velocity and angular velocity during the simulation. Implement keyboard controls or graphical sliders that directly change the robot's movement parameters during execution.

i. Extend the orbital mechanics program to simulate a Hohmann transfer between Earth and Mars using poliastro. Calculate the necessary delta-v and time for the interplanetary transfer. Additionally, plot the transfer trajectory and the orbits of Earth and Mars.

j. Compute the estimated engineering costs of a complex project over time using cost models implemented in Python. Assess different project options, and make a decision using present value analysis by passing then-year costs to the net present value routine.

4 Python Applications

Python can be used to create applications that run on local desktops, in browsers over the Internet or locally, on other devices, and within embedded systems. Interactive applications can be constructed with Python and supporting languages. Hardware and firmware may be necessary in other cases. The most popular technologies and platforms with Graphical User Interfaces (GUIs) are summarized in Table 4.1.

The frameworks and protocols for creating browser-based and desktop applications have associated libraries for interacting with Python scripts. Typically they work in conjunction with other code which is usually HTML, JavaScript, or framework-specific library functions and syntax. The Common Gateway Interface (CGI) is a web server protocol used by many languages. Pyoidide is a JavaScript implementation of Python that runs in the browser and PyScript (in development) is built on Pyodide. The web frameworks like Flask and Django include other components and tools with the Python libraries. Mobile devices are not covered here, but frameworks including Kivy are available. Simple examples and templates are described next.

Table 4.1

Technologies for Graphical User Interfaces

Technology	Description	Browser Internet	Browser Local	Desktop Application	Mobile Application
PyQT	Python library			✓	
tkinter	Python library			✓	
Kivy	Python library			✓	✓
Pyodide	JavaScript library	✓	✓		
PyScript	JavaScript library	✓	✓		
CGI	web server protocol	✓	✓		
Flask	web framework	✓	✓		
Django	web framework	✓	✓		

4.1 HTML WEB PAGES AND BROWSERS

HyperText Markup Language (HTML) is the standard markup language for web pages and applications. It defines the structure of a page, including its content, headings, paragraphs, links, images, forms and other logical elements. It is composed of elements defined by tags with less than and greater than signs <>. Tags usually come in pairs with opening and closing tags (e.g., `<head>` and `</head>`. Elements can have attributes, which provide additional information about the element. Attributes are specified within the opening tag as `<element attribute="value">`.

The "hello world" example in Listing 4.1 demonstrates a minimal template with content[1]. The basic elements are:

- `<!DOCTYPE html>` is a declaration specifying the document type to the browser.
- `<html lang="en">` represents the root of an HTML document and indicates the language used in the content.
- `<head>` contains meta-information about the document, such as character encoding, title, and stylesheets.
- `<body>` contains the visible content of the document.

The file would be saved with an .html extension and rendered when opened in a browser from either a local directory or Internet server URL.

Listing 4.1   HTML Template

```
<!DOCTYPE html>
<html lang="en">
<head>
<title>Hello Engineers</title>
</head>
<body>
Hello Engineers on the Internet
</body>
</html>
```

Python can print strings of HTML markup containing variables, expressions, and other dynamic content to be rendered by browsers. f-strings are a convenient way to generate content based on calculations. A basic example of dynamic web page creation with Python printing out HTML is in Listing 4.2. Scripts running on a web server will need the initial *shebang* (#!) line to identify the location of the interpreter for the server.

A multiline f-string with the `today` variable embedded in the HTML markup is then written to serve a dynamic web page showing the current date. Additional elements beyond the template in Listing 4.1 are the `<h3>` tag for a third level header and the hyperlink `<a href="test_report.html">Add Test Report</a>` pointing to the test report submission page in Listing 4.4. The HTML content is printed to the output, which is sent to the web browser to be displayed as a webpage. When opening the page in a browser, it will display as Figure 4.1.

This example shows an essential template to be used for HTML pages and how Python can write strings to generate any portions of a page. It can be a concatenation of multiple strings. There is no provision yet for user input to provide data or control the script execution. Other methods are described next that support data fields, interface controls and provisions for accessing them in Python.

---

[1]HTML 5 compliancy can always be checked at https://validator.w3.org/

Listing 4.2   HTML Web Page with Python

```python
#!/usr/bin/env python3

import datetime

title = "Engineering Test Portal"
today = datetime.datetime.now()
heading = f"Welcome to the Engineering Test Portal on ↵
    {today.strftime('%m-%d-%Y')}"

html_text = f"""
<!DOCTYPE html>
<html>
<head><title>{title}</title></head>
<body>
<h3>{heading}</h3>
<a href="test_report.html">Add Test Report</a>
</body>
</html>
"""
print(html_text)
```

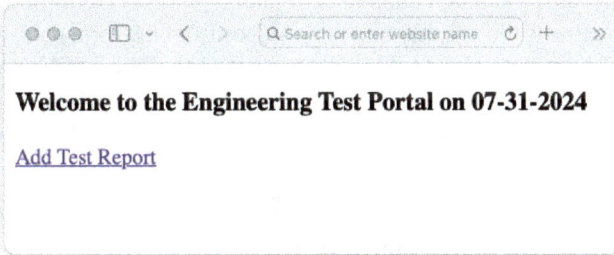

Welcome to the Engineering Test Portal on 07-31-2024

Add Test Report

Figure 4.1   Test Portal Web Page

These scripts need to run on a web server and be made executable[2]. When using the `cgi` module, a web server may need be configured to handle CGI. It comes as part of the predominant Apache web server.

4.2   BROWSER-BASED APPLICATIONS

There are a number of frameworks that can be used to create browser-based engineering applications ranging from data visualization tools to real-time simulations.

---

[2]E.g., with the command `chmod +x index.py`

Responsive user interfaces can be composed with HTML, CSS, and JavaScript with Python to handle the logic of the application. Collectively the markups and languages generate dynamic content that users can interact with in a web browser.

CGI performs server-side processing and will need a web server to run on the Internet or locally. Flask and Django have both offline and online capability. Flask is simpler and quicker to implement. Django has a steeper learning curve and is better suited for large scale website production and database applications. CGI requires the minimum extra steps beyond a standard Python program and interpreter.

4.2.1   CGI

CGI is a standard protocol for executing programs on a web server. With CGI, Python scripts can access inputs from HTML forms or externally to dynamically generate content. When a client makes an HTTP request for a CGI script, the web server passes the request to the Python interpreter to execute the script and returns the output to the client as the HTTP response.

A script accepts HTTP requests which may contain parameters for the script as typed in a URL, from input fields on the page, or from another program or web page. It returns an HTTP response as the output of the script. The `cgi` module has functions to parse incoming HTTP requests including handling form data and file uploads, debugging and error handling.

The kinetic energy calculator in Listing 4.3 demonstrates an interface form with two input fields and a submit button to calculate. It retrieves the form data using the `cgi` module, extracts the velocity and mass values from the form data, performs the calculations, and generates the HTML response with the calculated kinetic energy.

An HTML form is a document section containing controls for user input fields, checkboxes, radio buttons, etc. This input data is sent to a server for processing. Key elements in the form in Listing 4.3 are:

- `<form>` defines a form
- `action` attribute specifies the URL to send the form data
- `method` attribute specifies the HTTP method used to send the form data, usually POST or GET
- `<input>` defines an input field
- `type` attribute specifies the type of input field (text, number, submit, etc.)
- `name` attribute assigns a name to the input field
- `value` attribute sets the initial value of the input field
- `<label>` provides a description for form elements
- `<button>` creates a clickable button input.

A user enters values for velocity and mass in the input fields, and clicks the "Calculate" button to submit the form, The browser sends an HTTP request to the specified URL including the form data using the POST method. The form action calls itself to handle the input data and display the calculation result. This generalized example uses the environment variable `os.environ['SCRIPT_URI']` for the URL of the current script to call itself.

The `cgi.FieldStorage` class is used to parse the incoming form data. It extracts form field names and values from the request in order to access the submitted data. The `getvalue()` function is used to access data based on field names. The resulting web page is shown in Figure 4.2 after a calculation.

Listing 4.3   Kinetic Energy Calculator with CGI

```python
#!/usr/bin/env python3

import cgi
import os

# Retrieve form data (if any)
form = cgi.FieldStorage()
velocity = float(form.getvalue("velocity", "0"))
mass = float(form.getvalue("mass", "0"))
# Flag for Calculate button submitted
calculated = form.getvalue("calculated", False)

# Perform calculations when Calculate button is pressed
if calculated:
    kinetic_energy = .5* mass * velocity ** 2

# Print the HTML using an f-string
print(f"""
<!DOCTYPE html>
<html lang="en">
<head>
    <meta charset="UTF-8">
    <title>Kinetic Energy Calculator</title>
</head>
<body>
    <h3>Kinetic Energy Calculator</h3>
    <form action="{os.environ['SCRIPT_URI']}" method="post">
        <label for="velocity">Velocity (m/s):</label>
        <input type="number" name="velocity" value="{velocity}">
        </br>
        <label for="mass">Mass (kg):</label>
        <input type="number" name="mass" value="{mass}">
        </br>
        <button type=submit>Calculate</button>
        <input type="hidden" value="True" name="calculated"/></p>
    </form>
    Kinetic Energy: {kinetic_energy if calculated else ""} {"Joules"
        if calculated else ""}
</body>
</html>
""")
```

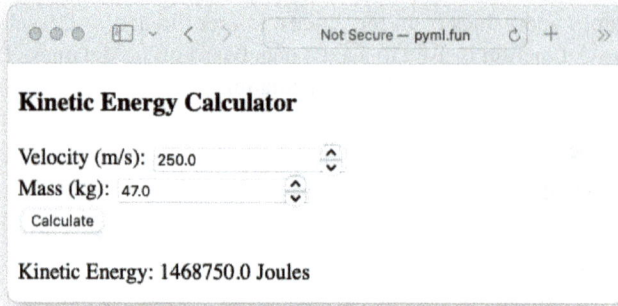

Figure 4.2   Kinetic Energy Calculator Application with CGI

The kinetic energy calculator has the input field names hardwired in. Adding more inputs the same way would be laborious and cumbersome. A generalized approach is desired to handle any number of fields. The `form.keys()` method returns a list of all the form field names which can be iterated over. This method for form handling is shown next.

The HTML in Listing 4.4[3] presents the test report page filled out in Figure 4.3. When submitted it sends form inputs to the script in Listing 4.5. The HTML line `<form action="save_data.py"method="post">` specifies the script filename to send the form data when it is submitted. The script processes the user data, writes it to a database, and replies with the web page in Figure 4.4.

Listing 4.4   HTML Page for Python Service

```html
<!DOCTYPE html>
<html lang="en">
<head>
    <meta charset="UTF-8">
    <meta name="viewport" content="width=device-width, ↙
        initial-scale=1.0">
    <title>Single Board Computer Test Report</title>
</head>
<body>
    <h1>Single Board Computer Test Report</h1>
    <form action="save_data.py" method="post">
    <b>Tester Name</b> <input type="text" name="name"><br>
        <table>
            <thead>
                <tr>
                    <th>Test</th>
```

---

[3]repeated form patterns are truncated

```
                    <th>Result</th>
                </tr>
            </thead>
            <tbody>
                <tr>
                    <td>Power Management & Boot Sequence</td>
                    <td>
                        <select name="power_boot_test">
                            <option value="passed">Passed</option>
                            <option value="failed">Failed</option>
                        </select>
                    </td>
                </tr>
                <tr>
                    <td>Processor & Memory Diagnostics</td>
                    <td>
                        <select name="processor_memory_test">
                            <option value="passed">Passed</option>
                            <option value="failed">Failed</option>
                        </select>
                    </td>
                </tr>
                ...
            </tbody>
        </table>

        <b>Additional Details</b><br>
        <textarea name="additional_details" rows="4"
            cols="50"></textarea><br>
        <br>
        <input type="submit" value="Submit Test Results">
    </form>
</body>
</html>
```

The processing of form data when the HTML page is submitted is done per Listing 4.5. It loops over `form.keys()` to get all the field names, and uses `form.getvalue(key)` to extract the values and add them to the data dictionary. The time entry to the data dictionary is explicitly written as another field.

It is straightforward to make responsive web applications combining CGI with JavaScript and CSS, but modern frameworks should be considered especially for high-traffic and data-intensive applications. CGI is a legacy technology and was recently planned to be deprecated in a future Python version. It isn't efficient for large data streams, but will remain suitable for many use cases and prevalent in existing implementations. However, it may become incompatible with other evolving libraries in the future that an application depends on.

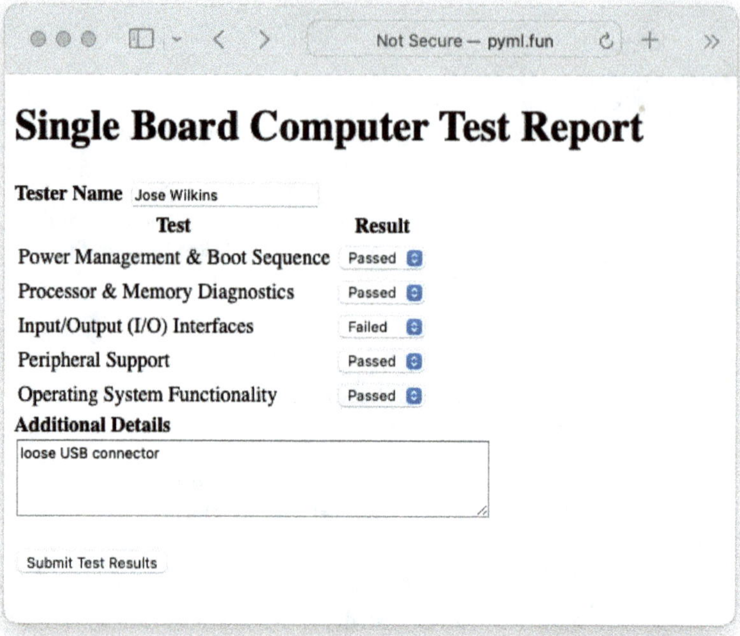

Figure 4.3   HTML Test Report Form for Python

Listing 4.5   CGI Form Service and Database Writing

```python
#!/usr/bin/env python3

import cgi
import csv
import os
import datetime

now = datetime.datetime.now().strftime("%Y-%m-%d %H:%M:%S")

print("Content-Type: text/html")
print()

form = cgi.FieldStorage()

data = {}
data["Time"] = now
for key in form.keys():
    data[key] = form.getvalue(key)
```

```python
csv_file = "test_report_data.csv"
file_exists = os.path.isfile(csv_file)

with open(csv_file, 'a', newline='') as csvfile:
    writer = csv.DictWriter(csvfile, fieldnames=data.keys())
    if not file_exists:
        writer.writeheader()
    writer.writerow(data)

print(f'<html><body><h3>Thank you {data["name"]}. The test report ↙
    was submitted successfully.</h3></body></html>')
```

Thank you Jose Wilkins. The test report was submitted successfully.

Figure 4.4   Python Test Report Response

4.2.2   FLASK

Flask is a lightweight and flexible web framework for building small to medium-sized web applications and APIs. It follows the WSGI (Web Server Gateway Interface) standard and uses Jinja2 for templating, allowing dynamic HTML generation. A Flask application typically has a directory structure to organize code and assets. At the top level is the main application file (e.g., app.py or main.py) and a directory called templates for HTML files. A static directory is for assets like CSS, JavaScript, and images as below.

```
/project-directory
    |-- app.py
    |-- /templates
    |    |-- index.html
    |    |-- layout.html
    |-- /static
    |    |-- style.css
    |    |-- script.js
    |-- requirements.txt
```

A Flask web application for a projectile calculator is in Listing 4.6. It simplifies the directory structure to a single Python file that writes HTML. It can be executed locally at the command line with the `python app.py` command, whereby it serves the web application at a local URL. It is initialized with `app = Flask(__name__)`, which creates a Flask application instance named app. The logic is set with

`@app.route('/', methods=['GET', 'POST'])`, which defines a route handler for the root URL that accepts both GET and POST requests. The `index` function handles requests for the route.

It retrieves the input from the form data using `request.form[]`, calls the calculation function, and renders `index_template` with calculated values for display. The resulting application window is shown in Figure 4.5.

In Flask and other templating systems, using the Jinja2 syntax `{{}}` is the recommended way to embed variables into the HTML string as is done for the input fields here. Flask templates offer additional functionality such as loops, conditionals, and filters that work with the `{{}}` syntax but not with f-strings.

Listing 4.6   Projectile Distance Calculator with Flask

```python
import numpy as np
from flask import Flask, render_template_string, request

app = Flask(__name__)

def calculate_distance(velocity, angle):
    g=9.81
    velocity = float(velocity)
    angle = float(angle)
    angle_rad = np.radians(angle)
    return (velocity**2) * np.sin(2 * angle_rad) / g

index_template = """
<!DOCTYPE html>
<html>
<head>
    <title>Projectile Distance Calculator</title>
</head>
<body>
    <h1>Projectile Distance Calculator</h1>
    <form method="post">
        <label for="velocity">Velocity (m/s):</label>
        <input type="text" name="velocity" id="velocity" value="{{ 
            velocity }}"><br><br>
        <label for="angle">Angle (degrees):</label>
        <input type="text" name="angle" id="angle" value="{{ angle 
            }}"><br><br>
        <input type="submit" value="Calculate">
    </form>
    <p>Distance: {{ distance }} meters</p>
</body>
</html>
"""

@app.route('/', methods=['GET', 'POST'])
def index():
```

```
    velocity = 0
    angle = 0
    distance = 0
    if request.method == 'POST':
        velocity = request.form['velocity']
        angle = request.form['angle']
        distance = f"{calculate_distance(velocity, angle):.1f}"
    return render_template_string(index_template, velocity=velocity, ⤶
        angle=angle, distance=distance)

if __name__ == '__main__':
    app.run(debug=True, port=0)
```

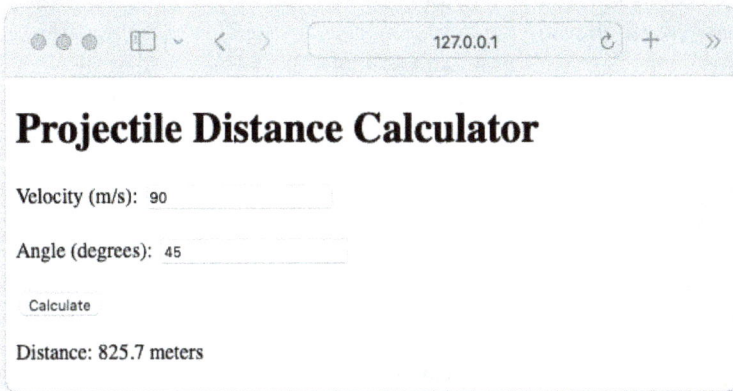

Figure 4.5   Projectile Calculator with Flask

4.2.3   PYODIDE

Pyodide enables Python to run in the browser by compiling CPython, the standard implementation of Python, to WebAssembly and embedding it in a JavaScript package. WebAssembly is a low-level, binary format optimized for fast browser execution. This allows for running Python on the client side in a web application instead of on a server. JavaScript functions can be called from Python and vice versa, enabling integration with existing web applications and JavaScript libraries. Pyodide provides access to the full Python standard library, but not all other libraries may be readily available in WebAssembly.

Pyodide can be used on any device with a modern web browser installed. The files can be hosted on an Internet server or used from local offline storage (though an Internet connection is required to access the online Pyodide script). Offline mode

is useful when a network connection is not available or reliable, or for easier local
development in a safe environment.

The example in Listing 4.6 is a projectile calculator application using Pyodide in
an HTML file with JavaScript . JavaScript handles user interaction and page manipulation while Python executes the calculations as a "separation of concerns". It relies
on the `pyodide.js` library to be read in for the Python interpreter. The JavaScript
allows asynchronous behavior with the `async` and `await` statements. The `async`
function defined as `main` allows for asynchronous operations without blocking the
main thread for Pyodide to execute the Python. The script adds a click event listener
to the Calculate button for user interaction. When clicked, it retrieves user input for
velocity and angle.

HTML is output, and the `main` function waits for button events from the click
event listener. The Python code is executed when the `Calculate` button is submitted. The code is sent to Pyodide using `pyodide.runPythonAsync` to run it asyn-
chronously while allowing the JavaScript code to continue execution. The output is
displayed on the page with JavaScript writing to the `result` field as shown in Figure
4.6. This example can be used as a basic template for data input fields, calculation,
and output display.

Listing 4.7   Projectile Calculator Web Application with Pyodide

```
<!doctype html>
<meta charset="utf-8">
<html lang="en">
<html>
<head>
    <title>Projectile Calculator</title>
    <script
        src="https://cdn.jsdelivr.net/pyodide/v0.20.0/full/pyodide.js">
    </script>
</head>
<body>
    <h1>Welcome to the Projectile Calculator</h1>
    <label for="velocity">Velocity (m/s):</label>
    <input type="text" id="velocity" style="width: 50px;">
    <br>
    <label for="angle">Angle (degrees):</label>
    <input type="text" id="angle" style="width: 50px;">
    <br>
    <button id="calculate_button">Calculate</button>
    <br>
    <p id="result"></p>

    <script type="text/javascript">
        // Asynchronous function allowing for non-blocking operations
        async function main() {
            // Load Pyodide and import the necessary packages
            let pyodide = await loadPyodide({
```

```
            indexURL: ∠
                'https://cdn.jsdelivr.net/pyodide/v0.20.0/full/'
        });
        await pyodide.loadPackage(['numpy']);

        // Add click event listener to the "Calculate" button
        document.getElementById("calculate_button").addEventListener
        ("click", function() {
            // Get the values of velocity and angle entered by the ∠
                user
            let velocity = ∠
                document.getElementById("velocity").value;
            let angle = document.getElementById("angle").value;

            // Run the Python code asynchronously using Pyodide
            pyodide.runPythonAsync(`
                import numpy as np

                # Define the Python function to calculate the ∠
                    distance
                def calculate_distance(velocity, angle):
                    g=9.81
                    velocity = float(velocity)
                    angle = float(angle)
                    angle_rad = np.radians(angle)
                    return (velocity**2) * np.sin(2 * angle_rad) / g

                # Call the calculate_distance function with user ∠
                    inputs
                distance = calculate_distance(${velocity}, ${angle})
                f"{distance:.1f}"
            `).then((result) => {
                // Display the calculated distance in the result ∠
                    paragraph
                document.getElementById("result").innerText = ∠
                    'Distance = ' + result + ' meters';
            });
        });
    }

    // Call the main function to start the application
    main();
</script>

</body>
</html>
```

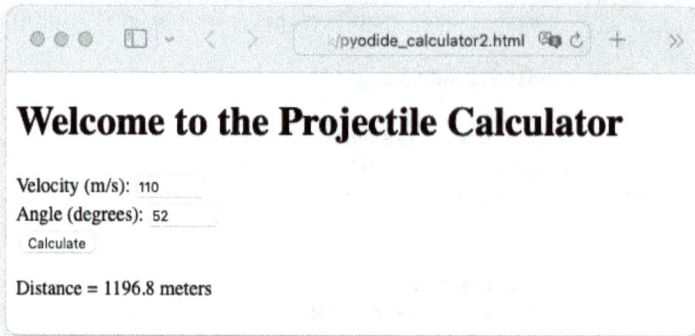

Figure 4.6   Projectile Calculator Web Application with Pyodide

4.2.4   PYSCRIPT

PyScript is a developing platform that uses Pyodide to enable almost pure Python without having to write JavaScript. It provides markup elements and widgets that are interspersed with HTML. Python can be either embedded within the web page and/or typed into code cells like a notebook. This makes it possible to create applications that are more interactive and dynamic than traditional HTML/CSS/JavaScript applications.

Listing 4.8 is an example of a simple notebook emulation with code cells implemented in an HTML file. The PyScript JavaScript, and CSS libraries must be identified in the HTML header like Pyodide (note that they can be copied to another Internet server or local storage, eliminating the online domain dependency). The markup tag py-repl delineates a code cell. The tag option auto-generate="true" will append a new cell below each cell that is run. This way one creates a dynamic notebook with multiple cells similar to a Jupyter Notebook. The initial cell is prepopulated with a print statement but could be left blank.

Listing 4.8   HTML File for Python Notebook with PyScript

```
<!DOCTYPE html>
<head>
  <link rel="stylesheet" ↵
      href="https://pyscript.net/latest/pyscript.css" />
  <script defer src="https://pyscript.net/latest/pyscript.js"></script>
</head>
<body>
  <py-repl auto-generate="true">
  print("Hello engineers around the world!")
  </py-repl>
</body>
</html>
```

A rendered web page is shown in Figure 4.7. Opening the file in a browser displays REPL cells similar to a Jupyter Notebook where one can write and execute (but not save) programs. Each code cell can be run separately. The initial pre-populated code cell has been run and additional code cells executed. This operates in playground mode and user edits cannot be saved without additional functions and server file access.

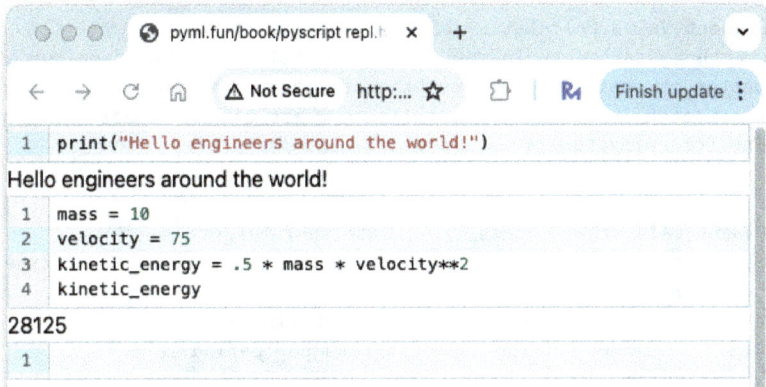

Figure 4.7   PyScript Notebook Example

PyScript with Pyodide can tap into the full Python ecosystem. Listing 4.9 scales up functionality with PyScript for an online present value calculator with graphics. The code from Listing 3.6 is placed in the script, which can then be called within a REPL cell. Explanatory library calls are put onto the web page for users with a calculated page shown in Figure 4.8.

Listing 4.9   Present Value Calculator Application with PyScript and Matplotlib

```
<!DOCTYPE html>
<html lang="en">
  <head>
    <meta charset="utf-8" />
    <meta name="viewport" ↵
        content="width=device-width,initial-scale=1" />

    <title>Net Present Value Plot</title>

    <link rel="icon" type="image/png" href="favicon.png" />
    <link rel="stylesheet" ↵
        href="https://pyscript.net/latest/pyscript.css" />
    <script defer ↵
        src="https://pyscript.net/latest/pyscript.js"></script>
    <style>
      body {padding: 10px;}
```

```
      .py-repl-run-button{opacity: .5}
    <style>
  </head>
  <py-config type="toml">
    packages = ["numpy", "matplotlib",]
  </py-config>
  <body>

<h5>Present Value Calculator</h5>
Enter Python statements in the code cells and click the green run ↵
    button or shift-enter to run the scripts. Additional scripts can ↵
    be created in the new code cells that are generated after a run. ↵
    The following functions are available:</br></br>

<p>
<b>present_value</b>(<i>cash_flow, rate, n</i>)</br>
Calculates the present value of a cash flow for given rate and time ↵
    period n.</br>
    &bull; <i>cash_flow</i>: A future cash flow value.</br>
    &bull; <i>rate</i>: The discount rate per period as a ↵
        fraction.</br>
    &bull; <i>n</i>: The future time period</br>
<p>
<b>net_present_value</b>(<i>cash_flows, rate, plot=True, ↵
    plot_pv=False, show_values=False</i>)</br>
Calculates net present value of cashflows and optionally plots them. ↵
    It returns a tuple containing the cash flow present values and ↵
    net present values for each period, and displays the plot.</br>

    &bull; <i>cash_flows</i>: A list of cash flows for each ↵
        period.</br>
    &bull; <i>rate</i>: The discount rate per period as a ↵
        fraction.</br>
    &bull; <i>plot</i>: Flag to display a bar plot of future values ↵
        and line for net present values (default is True) ↵
        (optional)</br>
    &bull; <i>plot_pv</i>: Flag to display present values on bar ↵
        plot (default is True) (optional)</br>
    &bull; <i>show_values</i>: Flag to display data values on the ↵
        bars and line (default is False) (optional)</br>
<p>
Cashflow inputs are defined as lists using the syntax ↵
    <code>cash_flows = [-30, 11, 21, 23, 34] </code>, where each ↵
    entry represents a period cash flow. The first entry is for time ↵
    zero with no discounting, the next entry is for period 1, etc.

<py-script>
import numpy as np
import matplotlib.pyplot as plt
```

```
import matplotlib as mpl
from matplotlib.ticker import MaxNLocator

mpl.rcParams['axes.spines.top'] = False
mpl.rcParams['axes.spines.right'] = False

def present_value(cash_flow, rate, n):
    return cash_flow / (1 + rate)**n
...
rest of present value listing per Section 3.3
...
</py-script>

<py-repl auto-generate='True'>
cash_flows = [0, 20, 23, 30]
annual_rate = .07
net_present_value(cash_flows, annual_rate, plot=True, plot_pv=True, ↙
    show_values=True)
</py-repl>


    </body>
</html>
```

4.3   DESKTOP APPLICATIONS

There are various Python frameworks that can be used to create desktop applications. They typically use the native GUI widgets of the operating system on which the application is running. PyQt is a cross-platform framework with an API for creating Qt-based applications. Qt is a widely used, cross-platform application framework for developing GUIs. Tkinter is a simple but powerful GUI framework that is built into the Python standard library. Kivy is a cross-platform GUI framework for creating mobile and desktop applications.

4.3.1   PYQT

A desktop PyQT program is in Listing 4.10. It can be run at the command line or made into a stand-alone application. It first imports requisite libraries for the application input fields, controls, and other graphic elements. The `ProjectileCalculator` class inherits from QWidget and defines the application's window. The `__init__` method initializes the window and calls `initUI` to set up the interface. The `initUI` method sets all the elements as described in the comments.

The `on_click` method is triggered by clicking the button, user input for velocity and angle is retrieved, and it calls the projectile function for calculations. It finally updates the corresponding labels with the calculated values formatted with one decimal place. The result of a calculation is in Figure 4.9.

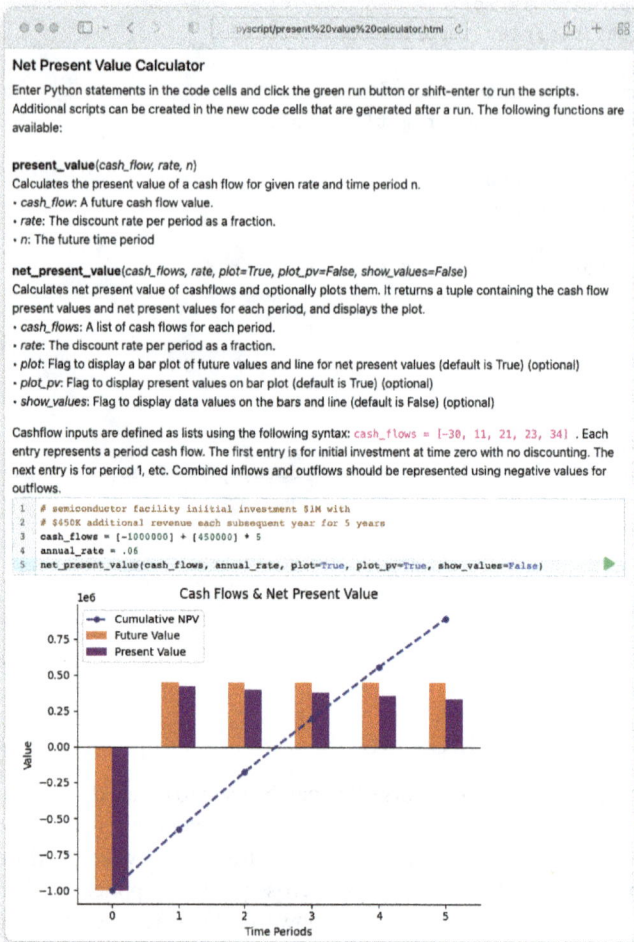

Figure 4.8   Present Value Calculator with PyScript and Matplotlib

Listing 4.10   Projectile Calculator Desktop Application with PyQt

```
import sys
from PyQt5.QtWidgets import QApplication, QWidget, QLabel, ↙
    QPushButton, QVBoxLayout, QLineEdit, QGridLayout
from PyQt5.QtGui import QFont
import math

def projectile(v0, angle):
    """ Returns the projectile flight time, maximum height and ↙
        distance given initial velocity in meters per second and ↙
```

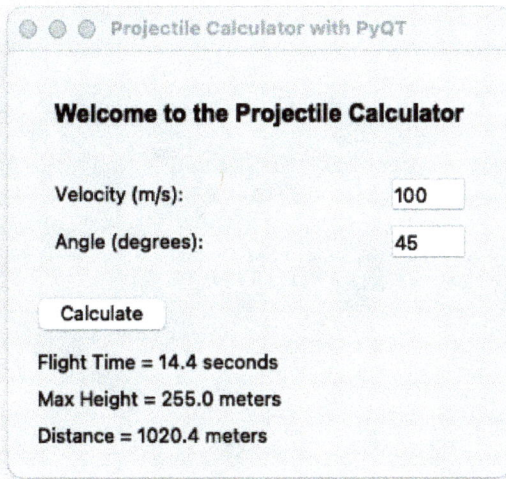

Figure 4.9   Projectile Calculator Desktop Application with PyQT

```
        launch angle in degrees. """

    g = 9.8# gravity (meters per second squared)
    angle_radians = 0.01745 * angle # convert degrees to radians
    flight_time = 2 * v0* math.sin(angle_radians) / g
    max_height = (1 / (2 * g)) * (v0 * math.sin(0.01745 * angle)) ** ↙
        2
    distance = 2 * v0** 2 / g * math.sin(angle_radians) * ↙
        math.cos(angle_radians)
    return(flight_time, max_height, distance)

class ProjectileCalculator(QWidget):
    """
    A PyQt-based GUI application for calculating and displaying the ↙
        trajectory of a projectile.

    This class creates a windowed application that allows the user ↙
        to input the initial velocity
    and launch angle of a projectile. Upon clicking the calculate ↙
        button, the application
    computes the projectile's flight time, maximum height, and ↙
        distance using the provided
    values. The results are then displayed in the window.
```

```
    Inherits:
        QWidget: Base class for all UI objects in PyQt.

    Methods:
        __init__(): Initializes the GUI components and sets up the ⤶
            application window.
        initUI(): Configures the UI elements, including labels, input ⤶
            fields, and buttons.
        on_click(): Handles the click event of the calculate button, ⤶
            performs the calculations,
                    and updates the result labels.

    Attributes:
        lbl (QLabel): A label that displays the title of the ⤶
            application.
        label_velocity (QLabel): A label for the velocity input field.
        line_edit_velocity (QLineEdit): A text field for the user to ⤶
            input the initial velocity.
        label_angle (QLabel): A label for the angle input field.
        line_edit_angle (QLineEdit): A text field for the user to ⤶
            input the launch angle.
        lbl_time (QLabel): A label to display the calculated flight ⤶
            time.
        lbl_height (QLabel): A label to display the calculated ⤶
            maximum height.
        lbl_distance (QLabel): A label to display the calculated ⤶
            distance.
    """

    def __init__(self):
        super().__init__()

        self.initUI()

    def initUI(self):
        self.setWindowTitle('Projectile Calculator with PyQT')
        self.setGeometry(300, 300, 250, 250) # Set initial position ⤶
            and size

        self.lbl = QLabel('Welcome to the Projectile Calculator', self)
        self.lbl.setFont(QFont('Arial', 16, QFont.Bold)) # Set the ⤶
            font to bold

        # Create the labels and text input fields for velocity and ⤶
            angle
        self.label_velocity = QLabel("Velocity (m/s):")
        self.line_edit_velocity = QLineEdit()
        self.line_edit_velocity.setFixedWidth(50) # Set a fixed width ⤶
            for the velocity input field
```

```python
        self.label_angle = QLabel("Angle (degrees):")
        self.line_edit_angle = QLineEdit()
        self.line_edit_angle.setFixedWidth(50) # Set a fixed width ↙
            for the angle input field

        # Create a container widget for the top label
        top_label_container = QWidget()
        top_label_layout = QVBoxLayout(top_label_container)
        top_label_layout.addWidget(self.lbl)

        # Layout the labels and text input fields for velocity and ↙
            angle
        layout = QGridLayout()
        layout.addWidget(self.label_velocity, 0, 0)
        layout.addWidget(self.line_edit_velocity, 0, 1)
        layout.addWidget(self.label_angle, 1, 0)
        layout.addWidget(self.line_edit_angle, 1, 1)

        # Set the layout for the widget
        container = QWidget(self)
        container.setLayout(layout)

        button = QPushButton('Calculate', self)
        button.setFixedWidth(100) # Set a fixed width for the button
        button.clicked.connect(self.on_click)

        self.lbl_time = QLabel(' ', self) # blank output label until ↙
            computed
        self.lbl_height = QLabel(' ', self) # blank output label ↙
            until computed
        self.lbl_distance = QLabel(' ', self) # blank output label ↙
            until computed

        vbox = QVBoxLayout()
        vbox.addWidget(top_label_container) # Add the container with ↙
            the top label
        vbox.addWidget(container) # Add the container with the labels ↙
            and text inputs
        vbox.addWidget(button)
        vbox.addWidget(self.lbl_time)
        vbox.addWidget(self.lbl_height)
        vbox.addWidget(self.lbl_distance)

        self.setLayout(vbox)

        self.show()

    def on_click(self):
```

```
        velocity = float(self.line_edit_velocity.text())
        angle = float(self.line_edit_angle.text())
        flight_time, max_height, distance = projectile(velocity, angle)
        self.lbl_time.setText(f'Flight Time = {flight_time:.1f} ↵
            seconds')
        self.lbl_height.setText(f'Max Height = {max_height:.1f} ↵
            meters')
        self.lbl_distance.setText(f'Distance = {distance:.1f} meters')


if __name__ == '__main__':
    app = QApplication(sys.argv)
    calculator = ProjectileCalculator()
    sys.exit(app.exec_())
```

4.4 EMBEDDED SYSTEMS

Python is used in a variety of embedded systems. It is suitable for example on single-board computers like the Raspberry Pi and BeagleBoard, where it can interact directly with hardware components through General Purpose Input/Output (GPIO) pins. This allows control of hardware to read sensor data and control many devices.

Python is also applicable in microcontroller-based systems, including platforms like Arduino when paired with MicroPython or CircuitPython. These implementations of Python are lightweight versions optimized to run on the limited resources typical of microcontrollers, enabling Python to control hardware like temperature sensors, motors, and communication modules.

Python can manage data collection and processing, communicate with cloud services for data logging, perform other web-based tasks, and even execute real-time hardware control. The selection of the following examples demonstrates Python as a practical tool for interfacing with hardware and developing embedded applications.

4.4.1 HARDWARE AND GPIO INTERFACE

Python has libraries to interact with standard GPIO (General Purpose Input/Output) pins on single-board computers like the Raspberry Pi and other hardware controllers. The GPIO pins are used to receive signals from sensors and other devices, or to control devices such as relays, small motors, or LEDs. These types of pins and their GPIO numbers are illustrated in the example pinout wiring diagram in Figure 4.10. The pin-out shows LEDs connected to pins 17 (blue) and 23 (orange) with an input button on pin 25 (red).

The GPIO libraries provide a software interface to read the state of an input pin or set the state of an output pin. When a program sets the state of an output pin, it sends a signal to the hardware that controls the voltage on the pin, either high or low. This signal can be used to control other devices that are connected to the pin, such as LEDs or motors. When the program reads the state of an input pin, it measures the

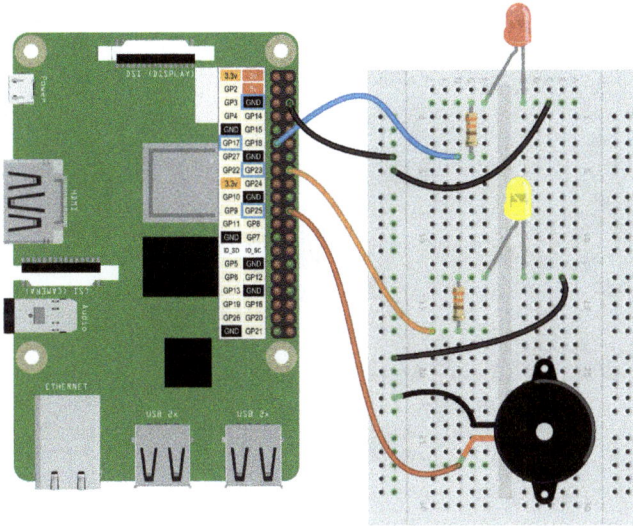

Figure 4.10   Example Raspberry Pi Wiring with GPIO Pins

voltage on the pin to determine whether it is high or low, which can indicate the state of a connected device, such as a button or sensor. They typically vary between 0 to 3.3 volts on a pin.

A program for a Raspberry Pi in Listing 4.11 controls LEDs per Figure 4.10 with button presses. It imports the RPi.GPIO library for Raspberry Pis, then sets the pin numbering mode for the physical pins on the board. One can then configure the desired pins as inputs or outputs, and use the library functions to read or write to the pins.

The main loop checks for button presses and toggles the blinking state. If blinking is active, it changes the LED state every second. When blinking is turned off, it immediately turns off all LEDs. A debounce mechanism in the toggle_blinking() function is implemented to prevent spurious multiple toggles from a single press.

Listing 4.11   Button Sensor Reading and LED Control with GPIO

```python
import RPi.GPIO as GPIO
import time

# Set the GPIO mode to BCM
GPIO.setmode(GPIO.BCM)

# Define the GPIO pins
BUTTON_PIN = 25 # Button input pin
LED_PINS = [17, 23] # LED output pins
```

```python
# Set up the button pin as an input with a pull-up resistor
GPIO.setup(BUTTON_PIN, GPIO.IN, pull_up_down=GPIO.PUD_UP)

# Set up the LED pins as outputs
for pin in LED_PINS:
    GPIO.setup(pin, GPIO.OUT)

# Global variables
blinking = False
last_toggle_time = 0

def toggle_blinking():
    global blinking, last_toggle_time
    current_time = time.time()
    if current_time - last_toggle_time > 0.3: # Debounce
        blinking = not blinking
        last_toggle_time = current_time
        if blinking:
            print("Starting LED blinking")
        else:
            print("Stopping LED blinking")
            for pin in LED_PINS:
                GPIO.output(pin, GPIO.LOW)

try:
    print("Press the button to toggle LED blinking (Ctrl+C to exit)")
    led_state = False
    last_blink_time = 0

    while True:
        current_time = time.time()

        # Check for button press
        if GPIO.input(BUTTON_PIN) == GPIO.LOW:
            toggle_blinking()

        # Handle LED blinking
        if blinking and current_time - last_blink_time >= 1:
            led_state = not led_state
            for pin in LED_PINS:
                GPIO.output(pin, led_state)
            last_blink_time = current_time

except KeyboardInterrupt:
    print("\nProgram stopped by user")

finally:
    GPIO.cleanup()
```

The next example in Listing 4.12 reads a variable potentiometer knob adjusted by a user to control the speed of a small motor using Pulse Width Modulation (PWM). It polls the voltages on the input pin and proportionally controls the motor according to the measured voltage. The voltage on the potentiometer pin is read and mapped to a speed value between using linear interpolation. The motor speed is then set using the ChangeDutyCycle function. The main loop repeats every 100 milliseconds.

Listing 4.12   Potentiometer Sensor Reading and Motor Control with GPIO

```python
import RPi.GPIO as GPIO
import time

# Set up GPIO pins
GPIO.setmode(GPIO.BCM)
input_pin = 17 # Potentiometer input
motor_pin = 22 # Motor control pin

GPIO.setup(input_pin, GPIO.IN)
GPIO.setup(motor_pin, GPIO.OUT)

# Set up PWM for motor control
motor_pwm = GPIO.PWM(motor_pin, 1000) # 1000 Hz frequency
motor_pwm.start(0) # Start with 0% duty cycle

# Constants for mapping voltage to motor speed
min_voltage = 0
max_voltage = 3.3
min_speed = 0
max_speed = 100

try:
    # Main loop to poll the voltage and adjust the motor speed
    while True:
        # Read analog voltage from potentiometer
        voltage = GPIO.input(input_pin) * max_voltage # fraction of ↙
            max voltage on pin

        # Map voltage to speed
        speed = (voltage - min_voltage) / (max_voltage - min_voltage) ↙
            * (max_speed - min_speed) + min_speed

        # Set motor speed
        motor_pwm.ChangeDutyCycle(speed)

        time.sleep(0.1 )

except KeyboardInterrupt:
    print("Program stopped by user")
```

```python
finally:
    motor_pwm.stop()
    GPIO.cleanup()
```

4.4.2   WEATHER STATION REPORTING

The script in Listing 4.13 worked as an actual weather station reporting service for several years. It periodically reads the current temperature on a local network web page written by a thermistor sensor program. The `urllib` module is used to open a URL for reading and sending as a GET request.

A dictionary is populated for the data update and encoded for the weather station API. It is encoded into `url_values` for the query string and placed after the ? which is the query string separator in `full_url`. The data dictionary keys become the parameter names, and the values become the parameter values in the query string. The `urllib.parse.urlencode()` function takes care of properly formatting and encoding the query string when making the HTTP request to the API.

Listing 4.13   Weather Station Update with API

```python
import time
import urllib.request
from time import gmtime, strftime

# Base URL for weather update API
url = 'http://rtupdate.wunderground.com/weatherstation/
        updateweatherstation.php'

while True:
    # Update every 60 seconds
    time.sleep(60.0)

    # Fetch current temperature from local network URL
    response = urllib.request.urlopen('http://192.168.1.178')
    temperature = response.read()

    # Data for API query parameters
    data = {
        'ID': 'KCASANDI115',
        'PASSWORD': 'xxx',
        'tempf': temperature,
        'action': 'updateraw',
        'dateutc': strftime("%Y-%m-%d %H:%M:%S", gmtime()),
        'realtime': 1,
        'rtfreq': 60.0,
        'softwaretype': 'custom python'
    }

    # Encode query string and append to base URL
```

```
url_values = urllib.parse.urlencode(data)
full_url = f"{url}?{url_values}"

# Send update as GET request with URL string
with urllib.request.urlopen(full_url) as response:
    print(f"Update sent. Response status: {response.status}")
```

4.4.3 DRONE FLIGHT CONTROL

Ardupilot open-source software is used to control many classes of vehicles. A Python script can be used to simulate a flight in Software In-The Loop (SITL) mode or be uploaded to an actual vehicle for autonomous flight control. See more details in the ArduPilot documentation at [1].

The simple auto-pilot script in Listing 4.14 connects to a drone quadcopter over a network connection, arms the vehicle, takes off to a target altitude of 10 meters, flies to a given waypoint, and lands the vehicle. An SITL simulation is launched from the command line or using the Ardupilot Mission Planner application built-in tools. The script is run from the command line which connects to SITL at the address and port tcp:127.0.0.1 :5762. The mission is displayed while executing in Mission Planner SITL mode per Figure 4.11.

For real hardware, a flight controller is connected to Mission Planner. The script is run on a ground control station or onboard computer connected to the flight controller. The entire flight is monitored in Mission Planner. This example would need to be modified for the Micro Air Vehicle Link (MAVLink) address for SITL or a real drone.

Listing 4.14 Drone Flight Control

```python
from dronekit import connect, VehicleMode, LocationGlobalRelative
import time

# Connect to the vehicle using MAVLink
vehicle = connect('tcp:127.0.0.1:5762', wait_ready=True)

# Arm the vehicle and set to guided mode
vehicle.mode = VehicleMode('GUIDED')
while not vehicle.is_armable:
    time.sleep(1)
vehicle.armed = True

# Take off to 10 meters altitude
target_altitude = 10
vehicle.simple_takeoff(target_altitude)
while True:
    altitude = vehicle.location.global_relative_frame.alt
    if altitude >= target_altitude * 0.95:
        break
```

Figure 4.11   Ardupilot Flight Simulation Map

```
    time.sleep(1)

# Fly to a waypoint
target_latitude = 33.2170
target_longitude = 117.3514
target_altitude = 20
target_location = LocationGlobalRelative(target_latitude, ↙
    target_longitude, target_altitude)
vehicle.simple_goto(target_location)

# Land the vehicle
vehicle.mode = VehicleMode('LAND')
while vehicle.armed:
    time.sleep(1)

# Disconnect from the vehicle
vehicle.close()
```

4.5 EXERCISES

1. Use one of the application platforms to modify the kinetic energy calculator. Add fields for user inputs such as height to calculate potential energy and gravitational force. Once the form is submitted, display both the kinetic and potential energy values, along with the gravitational force, on the interface.

2. Develop a simple application that takes inputs from users to solve a chosen engineering problem that produces multiple outputs or data streams. The application should display the results in the interface as tables and/or graphs.

3. Choose any example or exercise from Chapter 3, and implement it as a web-based, desktop, or mobile application using one of the application frameworks covered in this chapter.

4. Connect a physical sensor (e.g., DHT11 for temperature) to the GPIO pins of a Raspberry Pi, and read the data in real time using Python. Display the measurement on the console.

5. Use the GPIO library to implement PWM control of a device (e.g., for adjusting brightness or motor speed). Allow the user to input control levels (0 to 100%) through a command-line interface.

6. Develop a Python-based system to control multiple devices connected to GPIO pins. Use a button or numeric keypad to switch between controlling different devices. Each device may have its own control, such as a potentiometer for variable adjustments.

7. Implement a graphical user interface using Tkinter or PyQt that allows users to control multiple devices connected to GPIO pins. The current state of each device should be displayed in the interface.

8. Build a web-based interface using Flask to control GPIO-connected devices from a web browser. Display the current state of each device in the interface.

9. Create a data logger for an engineering experiment using one of the application platforms. Allow the user to input sensor data (e.g., temperature, pressure, velocity) through a form. The script should save these values to a CSV or Excel file and display a table of recent entries in the interface.

10. Build a web-based or desktop application that allows engineers to input variables for a Monte Carlo simulation (e.g., probability distributions, number of iterations). Once submitted, the script should run the simulation in Python and display a histogram of the results.

4.6 ADVANCED EXERCISES

1. Build a more complex web-based or desktop application that accepts user inputs and performs advanced engineering simulations with animated output (e.g., fluid dynamics or thermal analysis).

2. Create a web-based script that collects real-time sensor data from a laboratory setup and displays it on a web page. Collect the data using Raspberry Pis or other devices, ensuring real-time updates. This could serve as a simple web-based dashboard for monitoring lab experiments.

3. Implement a web service that interacts with physical hardware in real time. For example, create a web-based interface to control a robotic arm. The user should be able to input specific movements or coordinates, and the script should send commands to the hardware while receiving and displaying real-time feedback on the robot's status.

4. Create a DIY home or lab control system using Python to interface with real hardware and sensors. Utilize platforms like Raspberry Pi or other embedded systems. Assemble a control panel with buttons, knobs, or touchscreens to manage devices (e.g., lights or motors) and receive sensor data. Implement communication over WiFi, Ethernet, or Bluetooth as needed, and ensure reliable operation.

5. Design a Python program to monitor and log real-time data from multiple sensors on a Raspberry Pi. Implement a logging mechanism to store sensor data in a CSV file. Extend it to include an alert system that sends an email when a sensor value exceeds a predefined threshold. Use `threading` or `asyncio` to manage simultaneous sensor readings and ensure real-time data logging.

6. Implement a comprehensive weather station monitoring system using Flask or CGI. Create a dashboard that displays the real-time weather data on a webpage.

7. Adapt the drone control script, or create a new one, and test it in SITL mode. Ensure the drone successfully takes off, flies forward, hovers, and lands in the simulation. After validating the script in SITL, adapt it for use with a real drone, and connect via telemetry. Test the script with the real drone, ensuring it performs the same actions: takeoff, flight maneuvers, and landing. Be prepared to take manual control if the script behaves unexpectedly.

5 Processes and Tools

5.1 INTRODUCTION

The importance of well-defined engineering processes and mature tools cannot be overstated. A software process is commonly defined as a set of activities, methods, practices, and transformations that people use to develop software [10]. In an engineering context, a poor process may lead to a software defect producing a wrong homework answer or at worst causing a human disaster. Tools support processes through automation. They are a means to increase process efficiency, reduce defects, and eliminate rework for individuals and teams.

Applying processes and tools wisely will result in quality software that is cost-effective and meets its specified requirements. From undergraduates working on projects to large trans-disciplinary teams in industry, a robust process combined with appropriate tools can significantly enhance productivity, reduce errors, and ensure that the final system is reliable and maintainable (if necessary). This chapter overviews essential processes and tools to consider in Python environments. Another reference with comprehensive automation examples for individual workflows using Python is [21].

Usage of command line tools and scripting are introduced first as fundamental skills for automating tasks and integrating tools. These basic automation techniques are applied to processes in later sections including testing, debugging, version control, configuration management, documentation, performance optimization, and continuous integration and deployment.

5.2 COMMAND LINE TOOLS AND PYTHON SCRIPTING

Engineers should master the command line in conjunction with Python scripts as a powerful interface to interact with systems directly, automate tasks, and efficiently manage files. The Jupyter Notebook environment allows command lines in code cells for traditional shell commands and a rich set of magic commands provided by the Jupyter environment. These commands can simplify many common tasks. Command line operations in cells can be interspersed with Python scripts for seamless integration in a notebook and development workflow. They work on desktop and cloud-based environments.

For instance, the author utilized a combination of Google Colab notebooks, GitHub, Google Drive, and Overleaf to manage the developing content, figures, and diagrams and generate the book. These tools were integrated and supported with command-line scripts to automate regression testing of all scripts, version control, and batch production of all artifacts. This process not only ensured consistency across aspects of the book project but also significantly reduced the manual overhead associated with managing a large and link-rich document.

Shell commands are prefaced with a ! and provide broad system-level capabilities, interacting directly with the operating system. Magic commands are prefaced with a % or %% and are more integrated with Python. Shell commands are used for tasks outside the Python environment but can still interact with Python through variable passing and output capture. Knowing the command line is crucial for enhancing productivity. Many operations performed in a Python environment can be mirrored or expedited through command lines.

5.2.1 MAGIC COMMANDS

Magic commands in Jupyter Notebooks provide command line functionality to enhance the notebook's interactivity and functionality. They provide utilities specific to the Jupyter or IPython environment and run within the IPython kernel. Commonly used commands are in Table 5.1. They are designed to handle tasks such as setting modes, measuring execution time, or managing notebook metadata. Many commands such as ls, cd, mkdir, and rm are analogous to file management tasks common in Linux environments that can be used with Python.

Line magics are prefixed by a single percent sign (%) and apply to a single line of code. Commands like %ls to list directory contents and %pwd to show the current directory allow for quick file management within the notebook. Magics like %time and %timeit can be used to measure the execution time of code snippets.

Cell magics are prefixed by two percent signs (%%) and apply to an entire cell. For example, %%timeit will time the execution of all code in a cell.

Table 5.1

Common Magic Commands in Jupyter Notebooks

Syntax	Description
%ls	Lists files in the current directory.
%pwd	Prints the current working directory.
%cd <directory>	Changes the current working directory.
%time <command>	Times the execution of a single command.
%timeit <command>	Runs a command multiple times and reports the best time.
%run <script.py>	Runs an external Python script.
%matplotlib <mode>	Sets up Matplotlib to work interactively.
%%writefile <filename>	Writes the contents of the cell to a file.
%%capture	Captures the output of the cell and stores it in a variable.
%load <filename>	Loads code from an external script into the cell.
%reset	Resets the namespace by removing all variables.
%who	Lists all variables in the current namespace.

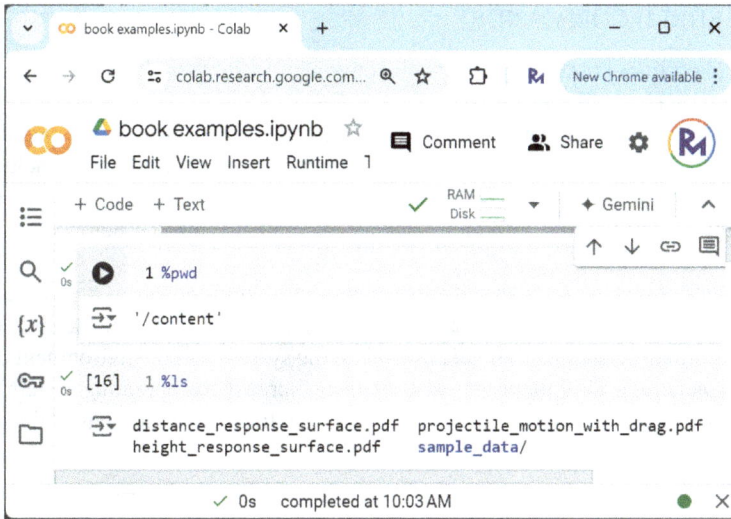

Figure 5.1   Magic Commands in Notebook

Examples of magic commands in a collapsed Google Colab notebook are in Figure 5.1. These were performed after running selected scripts to generate figures. The %pwd shows the current directory, which here is at the top level of the virtual file system. The command helps to confirm the directory context in which the notebook is operating, which is important to determine when working with relative file paths. The %ls command lists the contents of the directory as a list of files and subdirectories similar to Unix-like systems. It shows three PDF files that were generated and the default sample_data folder that is created in Colab notebooks.

Magic commands also include utilities for timing the execution of code. The %time and %timeit commands are used to measure how long a piece of code takes to run. The %time command returns the execution time of a single run of the code. For more rigorous performance analysis, the %timeit command runs it multiple times and provides statistical outputs.

The %timeit command is used next to compare execution times for the generation of one million random variates for delay time uniformly distributed between 0 and 10. It measures a list comprehension using the Python random.random function vs. the creation of a NumPy array. These timing commands are invaluable for performance optimization to identify and improve slow sections of code.

```
import random
%timeit random_delays = [random.random() * 10 for i in range(1000000)]
311 ms ± 8.67 ms per loop (mean ± std. dev. of 7 runs, 1 loop each)


import numpy
%timeit random_delays_np = np.random.rand(1000000) * 10
21 ms ± 4.68 ms per loop (mean ± std. dev. of 7 runs, 100 loops each)
```

5.2.2   SHELL COMMANDS

Table 5.2 outlines commonly used shell commands in Jupyter Notebooks, which can be used to perform file operations, manage directories, and run scripts directly from within a notebook cell. The ! character is used to execute shell commands directly from within a cell as if typed in a terminal. Common shell commands include !ls to list files in the directory, !pwd to print the working directory, or !pip install ∠ <package_name to install a Python package without leaving the notebook environment.

These commands interact with the operating system, enabling a wide range of tasks like file manipulation, package installation, and running system scripts. Unlike magic commands, shell commands execute in the system's shell environment outside the Python kernel. This allows any command available in the operating system's shell such as Windows or MacOS on desktops or Linux for many cloud-based IDEs.

Table 5.2

Common Shell Commands in Jupyter Notebooks

Syntax	Description
!ls	Lists files in the current directory.
!pwd	Prints the current working directory.
!cd <directory>	Changes the current working directory temporarily for the cell.
!mkdir <directory>	Creates a new directory.
!rm <file>	Removes a file.
!rmdir <directory>	Removes a directory.
!cp <source> <destination>	Copies a file from source to destination.
!mv <source> <destination>	Moves or renames a file or directory.
!echo <text>	Prints text to the output.
!cat <file>	Displays the content of a file.
!git <command>	Executes Git version control commands. E.g., !git clone, !git add <file>
!pip <command>	Executes pip commands for installing packages. E.g., !pip install numpy
!python <script.py>	Runs an external Python script using the Python interpreter.
!top	Displays system processes and resource usage.
!ps	Shows the currently running processes.
!history	Displays the command history.

The author uses the next commands in a Google Colab notebook for this book to populate data files to be read with Pandas scripts. The files are maintained in a public GitHub repository and cloned to the virtual filespace with git clone into a directory with the same name. The command cp -r copies them (r for recursively if there are subdirectories) into the current top level directory. The command `ls` lists the filenames showing they are ready for usage. Alternatively they can be viewed in the Colab file view in Figure 5.2.

```
!git clone ↙
    https://github.com/madachy/what-every-engineer-should-know-about-
    python-data-files
!cp -r what-every-engineer-should-know-about-python-data-files/*.* .
!ls

'air densities.xlsx'    'features and testing hours regression.xlsx'
boards.csv       sample_data
'drone cost weight data rate.csv'    turbine_data.csv
engineering_data.csv      what-every-engineer-should-know-about-python
-data-files    'features and testing hours.csv'
```

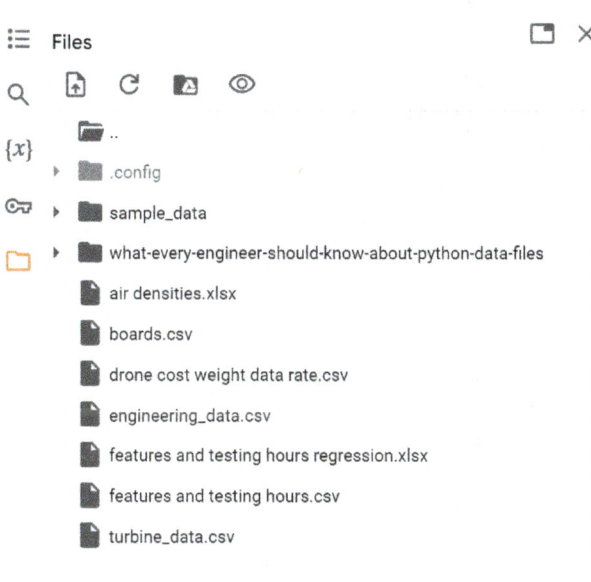

Figure 5.2   File List after Shell Commands

The shell syntax also supports the combination of Python variables with shell commands. The following example in Listing 5.1 is run at the beginning of a notebook that automates the batch running of multiple scripts and generation of figure files. It sets style defaults, connects to Google Drive, and defines a writing function. A Python variable is used to specify a filename for writing and passed to the `cp` shell command. This integration of Python and shell commands within a single environment is a key strength of Jupyter Notebooks to leverage both the power of Python and the flexibility of the command line.

Listing 5.1   Notebook Automation with Python and Shell Command Integration

```python
import matplotlib as mpl
import matplotlib.pyplot as plt

# plot defaults
mpl.rcParams['axes.spines.top'] = False
mpl.rcParams['axes.spines.right'] = False
mpl.rcParams['font.size'] = 9
# standard figure resized 75%
mpl.rcParams['figure.figsize'] = (.75 * 4.6, .75 * 3.45)
mpl.rcParams['mathtext.fontset'] = 'stix' #cm
mpl.rcParams['font.family'] = 'STIXGeneral'

# mount google drive to copy files
from google.colab import drive
drive.mount('/gdrive')

# write Matplotlib figure files after scripts
def write_book_figure(file_name):
    plt.savefig(f"{file_name}.pdf")
    !cp -r '{file_name}.pdf' '/gdrive/My Drive/What Every Engineer ↙
        Should Know About Python/figures/{file_name}.pdf'
```

The filename stored in `file_name` is passed to the `cp` command, allowing the shell command to dynamically operate on files based on the notebook's Python logic.

5.2.3   PURE PYTHON

Python-only scripts can also perform many of the operations of command lines. This can be advantageous in some contexts. It is important to use magic and shell commands judiciously. For example, if portability across different operating systems is important, relying too heavily on shell commands could be problematic since the commands may not behave consistently. In this case it is recommended to use cross-platform Python code for tasks that might otherwise be handled by shell commands.

As examples, a Python script can be used to batch rename files, automate backups, or even deploy applications. The script in Listing 5.2 automates the backup of project files. It uses Python's `os` and `shutil` libraries to copy all files from a source directory to a backup directory.

Listing 5.2   Project File Backup

```python
import os
import shutil

def backup_project(source_dir, backup_dir):
    if not os.path.exists(backup_dir):
        os.makedirs(backup_dir)
    for filename in os.listdir(source_dir):
```

```
        file_path = os.path.join(source_dir, filename)
        shutil.copy(file_path, backup_dir)
    print(f"Backup completed to {backup_dir}")

backup_project('/project_dir', '/backup_dir')
```

5.3   ITERATIVE DEVELOPMENT

One should iteratively "build a little, test a little" in increments to ensure a program works while making it easier to detect and fix errors. Writing more code before testing and debugging increases the chance of errors and makes it harder to find them. Iterative and incremental development reduces the complexity of debugging by focusing on small, manageable changes.

Try to start with a program that is already known to work correctly. This could be an example from a book or a previously developed program similar to the one being created. If such a program is not available, begin with a simple, correct one. Run the initial program to verify that it passes basic functional tests and behaves as expected.

In early iterations, stubs can be used in place of functions to mimic outputs in order to test overall logic. They are then incrementally replaced with the functions. Constants can be used to represent random variables until they are formulated in later increments.

Implement a small change at a time that can be immediately tested. A change is testable if it produces an observable outcome or effect that can be easily checked against expected results. The expected result should be known beforehand, or it should be possible to verify it through an oracle, alternative calculation, or comparison.

Execute and verify the program after making each change. Determine if it produces the expected output. If the change is successful, proceed with the next modification. Debugging may be necessary; otherwise determine the source. By keeping changes small, identifying the cause of a problem should be straightforward and less time-consuming.

Scaffolding code may help in testing to produce output that can be verified. This temporary code is used during development to test the program and is removed once it is no longer needed. While adding scaffolding may seem like an extra step, it often saves time by reducing the effort required for debugging.

5.4   TESTING AND DEBUGGING

Testing and debugging are fundamental practices to ensure code reliability and correctness. Automated testing is important to create repeatable and reliable tests that can be run frequently to catch defects early. Unit tests should be designed to validate the functionality of individual components and should be automated to save time and reduce errors. Debugging, on the other hand, involves identifying and fixing defects, often using tools like pdb to step through code and examine the state of a program during execution. The combination of these practices ensures that software behaves as expected and that any issues can be promptly addressed.

5.4.1 AUTOMATED TESTING

Automated testing is a fundamental practice in software development that ensures the reliability and correctness of code. By defining tests that cover various aspects of the codebase, developers can catch bugs early, reduce the risk of regressions, and maintain a high level of code quality throughout the development lifecycle. Automated tests can be run frequently, providing continuous feedback on the software quality.

Testing frameworks such as `unittest` and `pytest` are commonly used in Python to write and execute automated tests. These frameworks support the creation of test cases for individual functions, classes, and modules, as well as integration tests that verify the behavior of multiple components working together. By incorporating automated testing into the development workflow, software quality can be maintained as the code evolves.

A test harness is a collection of automated tests that are executed together to validate the behavior of a codebase. In Python, a test harness can be constructed using frameworks like `unittest` or `pytest`, which provide tools for defining and running test cases. Test cases should be defined for all expected input combinations.

The following example uses the `unittest` framework to test the present value program in Section 3.6. It verifies two test cases by comparing the final net present value against expected results computed independently. The function `npv` calls the `present_value` function from Listing 3.6 and returns the summary.

An assertion using the `assertAlmostEqual` method checked that the results match within a delta of one unit. The delta is chosen based on required accuracy and tolerance of small rounding errors. An example of successful run is shown.

Listing 5.3   Unit Test for Present Value Program

```python
import unittest

def npv(cash_flows, annual_rate, n=12):
    """
    Calculates the net present value of a series of cash flows with ↙
        compounding.
    """
    effective_rate = annual_rate/n
    return sum(present_value(cash_flow, effective_rate, time) for ↙
        time, cash_flow in enumerate(cash_flows))

class TestNPVFunction(unittest.TestCase):

    def test_npv(self):
        test_cases = [
            {
```

```
            "cash_flows": [-100000] + [0] * 23 + [50000],
            "annual_rate": 12 * .0075,
            "n": 12,
            "expected": -58208
        },
        {
            "cash_flows": [0, 20, 23, 30],
            "annual_rate": .07,
            "n": 1,
            "expected": 18.69 + 20.09 + 24.49
        }
        # Add more test cases here as dictionaries
    ]

    for case in test_cases:
        with self.subTest(case=case):
            result = npv(case["cash_flows"], case["annual_rate"], ⤸
                case["n"])
            self.assertAlmostEqual(result, case["expected"], ⤸
                delta=1) #

suite = unittest.TestLoader().loadTestsFromTestCase(TestNPVFunction)
unittest.TextTestRunner().run(suite)

----------------------------------------------------------------------
Ran 1 test in 0.015s

OK
<unittest.runner.TextTestResult run=1 errors=0 failures=0>
```

Running this test harness after each update to the present_value function ensures that any changes do not introduce errors. Suppose the last test result was miscalculated. The output report would look like the following. It flags the erroneous results and amount deviation from expected.

```
======================================================================
FAIL: test_npv (__main__.TestNPVFunction) (case={'cash_flows': [0, 20,
23, 30], 'annual_rate': 0.07, 'n': 1, 'expected': 65.27})
----------------------------------------------------------------------
Traceback (most recent call last):
  File "<ipython-input-5-b2f0754dec25>", line 32, in test_npv
    self.assertAlmostEqual(result, case["expected"], delta=1)  #
AssertionError: 63.2696 != 65.27 within 1 delta (2.0003 difference)


----------------------------------------------------------------------
Ran 1 test in 0.001s

FAILED (failures=1)
<unittest.runner.TextTestResult run=1 errors=0 failures=1>
```

This example only checked a single output parameter for each test case. Multiple parameters and entire data structures can also be compared. A more thorough and rigorous test harness would check the output lists of present values and cumulative net present values for different sequence lengths.

5.4.2 DEBUGGING TECHNIQUES

Debugging is the process of identifying and fixing errors in software. It is an essential skill as even well-written code can contain bugs. Finding them can be a challenge. Python provides several tools that make debugging easier by setting breakpoints, stepping through code, inspecting variables during execution, and understanding the flow of execution.

The `pdb` module is Python's built-in debugger. Adding the following `import` and `pdb.set_trace()` at the function beginning will pause execution and the debugger entered there. It goes into a command mode accepting debugger commands. They can be executed to inspect the current state of the program, step through the code line by line, and evaluate expressions. This interactive approach helps identify where and why errors occur more quickly than adding print statements throughout.

```python
import pdb

def projectile(v0, angle):
    """ Returns the projectile flight time, maximum height and ↙
        distance given initial velocity in meters per second and ↙
        launch angle in degrees. """
    pdb.set_trace()
    g = 9.8# gravity (meters per second squared)
    angle_radians = 0.01745 * angle # convert degrees to radians
    flight_time = 2 * v0* sin(angle_radians) / g
    max_height = (1 / (2 * g)) * (v0 * math.sin(0.01745 * angle)) ** ↙
        2
    distance = 2 * v0** 2 / g * sin(angle_radians) * cos(angle_radians)
    return(flight_time, max_height, distance)

projectile(55, 45)
```

The following debugging session shows it pauses at line 6 to enter the debugger. Commands are then provided starting with `help` that shows the list of available commands, the a command to print the argument list of the current function, and `step` to step to the next line 7 where the next command can be specified.

```
> <ipython-input-4-7fd730a5c1e3>(6)projectile()
      4       """ Returns the projectile flight time, maximum height ↙
              and distance given initial velocity in meters per second ↙
              and launch angle in degrees. """
      5       pdb.set_trace()
----> 6       g = 9.8# gravity (meters per second squared)
      7       angle_radians = 0.01745 * angle # convert degrees to radians
```

```
      8       flight_time = 2 * v0* sin(angle_radians) / g

ipdb> help

Documented commands (type help <topic>):
==========================================
EOF     commands  enable    ll        pp       s             until
a       condition exit      longlist  psource  skip_hidden   up
alias   cont      h         n         q        skip_predicates w
args    context   help      next      quit     source        whatis
b       continue  ignore    p         r        step          where
break   d         interact  pdef      restart  tbreak
bt      debug     j         pdoc      return   u
c       disable   jump      pfile     retval   unalias
cl      display   l         pinfo     run      undisplay
clear   down      list      pinfo2    rv       unt

Miscellaneous help topics:
===========================
exec  pdb

ipdb> a
v0 = 55
angle = 45
ipdb> step
> <ipython-input-4-7fd730a5c1e3>(7)projectile()
      5       pdb.set_trace()
      6       g = 9.8# gravity (meters per second squared)
----> 7       angle_radians = 0.01745 * angle # convert degrees to radians
      8       flight_time = 2 * v0* sin(angle_radians) / g
      9       max_height = (1 / (2 * g)) * (v0 * math.sin(0.01745 * ⤸
              angle)) ** 2
```

Other debugging tools such as Jupyter Notebook's built-in features or PyCharm's integrated debugger provide graphical interfaces for setting breakpoints, inspecting variables, and controlling execution flow. These tools enhance the debugging process by making it easier to visualize and navigate through complex code.

Effective debugging may also include adding print statements to output the values of variables, using logging to record program execution, and writing automated tests to isolate and reproduce bugs. Combining these techniques with debugging tools helps identify defects and resolve them to ensure reliable software.

5.5   USING GENERATIVE AI

Generative AI can provide immense leverage for engineers, enabling them to work at a higher cognitive level without getting lost in the details of syntax and repetitive tasks. By allowing AI to handle more routine aspects such as typing initial code, maintaining correct indentation and syntax, and performing tedious data reformatting, engineers can focus on higher-level design and problem-solving tasks. This approach facilitates more efficient integration of solutions with greater confidence.

To maximize the benefits of AI tools, engineers should decompose large, complex tasks into smaller, manageable parts that AI can handle with minimal uncertainties. This strategy reduces the risk of errors and simplifies the integration of generated solutions. As task complexity increases, it becomes far riskier for AI to generate reliable solutions.

The use of AI tools can alleviate some of the burden of syntax memorization. Engineers no longer need to remember every specific detail of Python syntax as they can describe their intent in unambiguous natural language. For the best results, it is still beneficial to lace low-level syntax into the queries. This helps the AI better understand the task at hand and reduces the likelihood of generating substandard code.

Generative AI is not a silver bullet though, and there are limitations and pitfalls to be aware of. AI solutions are unlikely to be trained specifically on a unique problem unless it is a simple toy problem. Never expect AI to provide a 100% defect-free solution right out of the box. Generated solutions will have correct syntax but may fail to execute properly due to logical flaws. Observed flaws include missing necessary library imports, using outdated or non-existent library functions, generating overly complex solutions, or omitting crucial steps in logic flow.

Therefore, do not blindly trust AI outputs and always verify them. Engineers must focus on evaluating and testing AI-generated solutions. Understanding, inspecting, and critiquing the provided code is essential for identifying when a solution might be fundamentally flawed, overly complex, or missing a nuance for an application, or when it might be more efficient to discard it and start over.

To effectively use AI tools, engineers should approach them as part of a larger toolkit that complements their understanding of Python and software development principles. A strong grasp of Python terms, concepts, and syntax remains essential for utilizing AI effectively and understanding the generated code. Being able to understand and critique the provided code is essential. Sometimes, it may be evident that a solution is unfeasible, obviating the need for testing. AI solutions may require complete discarding or substantial refinement.

Rigorous testing and verification are critical. Always test AI-generated code before moving to the next iteration, integrating it into a larger project, or deploying it for use. Automated testing should be employed to verify the functionality of AI-generated code, ensuring that it produces the correct results under various conditions. Spend time upfront to define expected results. Additionally, code reviews should be conducted to assess the readability, maintainability, and performance of the generated code. Careful inspection will ensure it meets requirements and adheres to best practices.

Expect to refine or refactor AI-generated solutions. Even in the best cases, some adjustments will likely be needed to optimize the code for specific requirements or to improve readability and maintainability. The benefits of generative AI can only be realized after gaining a solid understanding of the fundamentals and the ability to manually program.

Define data structures upfront instead of relying on AI-generated ones, which can be risky. Start with a clear definition based on the problem domain. Specify appropriate data structures for tasks by considering the nature of the data and the operations required. For instance, lists are best suited for ordered homogeneous data, while dictionaries are better for key-value pairs where fast lookups are needed. For many engineering applications, starting with user-defined dictionaries can be productive, as they frame a problem with natural terms related to the domain, making the implementation more intuitive.

Maintain a balanced approach between generative AI and human thought. Avoid becoming overly dependent on AI tools. While they can accelerate development, they should not replace a deep understanding of the problem or the principles of software development. Spending time formulating and refining a query to be very specific before presenting it to an AI tool is a worthwhile investment.

Overall, generative AI tools can significantly enhance efficiency, but they should be used thoughtfully and strategically. Understanding the underlying concepts, rigorously testing AI solutions, and maintaining a balance between leveraging AI assistance and applying one's own problem-solving skills are crucial for crafting robust and efficient solutions. Some of these principles are demonstrated in the following case study example.

5.5.1   CASE STUDY

In this case study example, the author iteratively developed a parametric systems engineering cost model with generative AI (ChatGPT). The seven iterations detailed in this section demonstrate the principles of iterative development, automated testing, structuring of AI queries, and assessing the generated solutions. This application is the Constructive Systems Engineering Cost Model (COSYSMO) and the resulting program available in the *se-lib* library at https://github.com/se-lib/se-lib/blob/main/selib/cost_models/cosysmo.py.

The equation to implement is:

$$Effort = A * Size^E * \prod_{i=1}^{n} EM_i \qquad (5.1)$$

where
  Effort is measured in Person-Months
  Size is a weighted sum of requirements, interfaces, algorithms, and scenarios
  A is a calibration constant derived from project data
  E represents a diseconomy of scale
  EM_i is the effort multiplier for the i_{th} cost driver.
The geometric product $\prod EM_i$ is called the Effort Adjustment Factor (*EAF*) that is multiplicative to the nominal effort.. The overall effort is then decomposed by phase and activities into constituent portions using average percentages.

The application was built from the inside-out beginning with a main equation, incrementally replacing stubbed out parameters with more complex calculations. ChatGPT generated new functions and populated data structures, while the author integrated its outputs and performed testing. The author is familiar with the application and planned the increments based on experience.

Preparation prior to starting the chat session included getting armed with the following:

- implementation of the single line core equation to be built around (not shown to ChatGPT)

- a unit test example as a template to modify

- reference test cases with expected results to test incremental capabilities

- documents with tables of parameter values to copy

- desired nested dictionary structures for cost factors and weight parameters

The approach can be categorized as "test-driven development" because the testing framework was created first. It was also updated for more detailed test cases before more complex code was integrated. The final, consolidated script is in Listing 5.4. The generated functions were modified for clarity during integration.

The first iteration develops a unit test class that will be used to test successive iterations of the main application. It starts with a known test harness used for the present value program shown previously in Section 5.4.1.

For the first tests, the `size` parameter is an aggregate rolled up value and `eaf` is a nominal value. They stand for the outputs of functions to create next and will be replaced with more inputs when the functions are incorporated.

1. Generate Test Harness with Defined Test Cases

```
Q. Use this test case class as a template to modify for testing a new
function "cosysmo" that returns a value for "effort" and name it
"TestCOSYSMOFunction". Modify it for the following test case inputs and
expected results.  Check that the results match within 1 decimal place:
1) "size": 628, "eaf": 1.0,  "expected_effort": 235
2) "size": 628, "eaf": .893,  "expected_effort": 209.9

class TestNPVFunction(unittest.TestCase):

    def test_npv(self):
        test_cases = [
            {
                "cash_flows": [-100000] + [0] * 23 + [50000],
                "annual_rate": 12 * .0075,
                "n": 12,
                "expected": -58208
    ...
```

2. Create Cost Factor Dictionary

In this step the structure of a cost factor dictionary is explicitly defined with examples. The dictionary will be needed to calculate the `eaf` parameter currently stubbed at its nominal value. It is shown as `cost_factors_dict` in Listing 5.4.

```
Q. create a dictionary keyed by cost factor names and the values are
dictionaries containing ratings and effort multipliers. an example for
two factors is below:
cost_factors_dict = {"rely": {"Very_Low": 0.82, "Low": 0.92, "Nominal": 1.0},
 "cplx": {"Very_Low": 0.73, "Low": 0.87}}
```

```
populate the new dictionary with the cost factor names, ratings and effort
multipliers copied below.
        Very Low    Low    Nominal    High    Very High    Extra High
Requirements Understanding    1.87    1.37    1.00    0.77    0.60
Architecture Understanding    1.64    1.28    1.00    0.81    0.65
Level of Service Requirements    0.62    0.79    1.00    1.36    1.85
Migration Complexity        1.00    1.25    1.55    1.93
Technology Risk    0.67    0.82    1.00    1.32    1.75
...
```

3. Write EAF Function

The function to calculate eaf using the cost factor dictionary is specified. Model default values are specified for factors when they aren't set by the user.

```
Q. now write a function "eaf" that receives a dictionary of ratings,
e.g. {"Requirements Understanding": "Low", "Migration Complexity": "High",
...} and returns the output "eaf" which stands for Effort Adjustment Factor.
It shall loop through the cost factors and multiply the effort multipliers
associated with their ratings to calculate "eaf". Assume a default rating
for any non-provided cost factor to be "Nominal".
```

The eaf function is integrated with the existing code to replace the stubbed value of 1. It is ready for testing to be set up in the next step.

4. Update Test Cases for Cost Factor Dictionary Input

The test cases were updated for more detailed inputs to test the eaf stub replaced by the full function.

```
Q. now modify this class so the "eaf" test case inputs are replaced by the
dictionary inputs. the first case shall be a blank dictionary (with all
cost factors defaulted to nominal) and the second case uses the
cosysmo_case_study_2 dictionary input from above:
class TestCOSYSMOFunction(unittest.TestCase):

    def test_cosysmo(self):
...
```

5. Phase Effort Function

```
Q. write a function "phase_effort" that will decompose the cosysmo effort
into these constituent activities and percentages of the total effort.
have it receive "effort", return a dictionary of the sub-effort and print
a table of the effort outputs:
Activities Effort %
Acquisition and Supply 7%
Technical Management 17%
```

```
System Design 30%
Product Realization 15%
Technical Evaluation 31%
```

6. Create Size Weight Nested Dictionaries

A table was copied from a PDF reference and pasted into the chat to populate the dictionary `size_weights`.

```
Q. Below is a table of COSYSMO Size Drivers and size weights by complexity.
Populate a dictionary of dictionaries "size_weights" keyed by the driver
names and the values are dictionaries keyed by the complexities and the
values are the weights:
Driver      Easy    Nominal    Difficult
System Requirements    0.5    1.00    5.0
Interfaces    1.1    2.8    6.3
Critical Algorithms    2.2    4.1    11.5
Operational Scenarios    6.2    14.4    30
```

7. Total Size Function Using Weights

The last step to complete the full model is to use the size weights to computer the total aggregate size. A detailed cost estimate can be generated with this last function integrated.

```
Q. write a function "total_size" that receives a dictionary of dictionaries
for size driver counts like {"System Requirements": {"Easy": 12, "Nominal":
3, ...} and returns the total size by multiplying the counts by weights.
e.g. for this example, 12*0.5 + 3*1.0 etc.
```

Listing 5.4   COSYSMO Cost Model

```python
def cosysmo(size_drivers, cost_factors):
    return .254 * eaf(cost_factors) *total_size(size_drivers)**1.06


def eaf(ratings_dict):
    effort_adjustment_factor = 1.0

    for cost_factor, rating_values in cost_factors_dict.items():
        # Get the rating from the ratings_dict or default to ↙
            'Nominal' if not provided
        rating = ratings_dict.get(cost_factor, "Nominal")

        # Retrieve the effort multiplier for the rating
        multiplier = rating_values.get(rating, 1.0) # default to 1.0 ↙
            if the rating doesn't exist for some reason

        # Multiply EAF by each effort multiplier
        effort_adjustment_factor *= multiplier

    return effort_adjustment_factor
```

```python
cost_factors_dict = {
    "Requirements Understanding": {"Very_Low": 1.87, "Low": 1.37, ↙
        "Nominal": 1.00, "High": 0.77, "Very_High": 0.60},
    "Architecture Understanding": {"Very_Low": 1.64, "Low": 1.28, ↙
        "Nominal": 1.00, "High": 0.81, "Very_High": 0.65},
    "Level of Service Requirements": {"Very_Low": 0.62, "Low": 0.79, ↙
        "Nominal": 1.00, "High": 1.36, "Very_High": 1.85},
    "Migration Complexity": {"Nominal": 1.00, "High": 1.25, ↙
        "Very_High": 1.55, "Extra_High": 1.93},
    "Technology Risk": {"Very_Low": 0.67, "Low": 0.82, "Nominal": 1.0↙
        0, "High": 1.32, "Very_High": 1.75},
    "Documentation": {"Very_Low": 0.78, "Low": 0.88, "Nominal": 1.00↙
        , "High": 1.1 3, "Very_High": 1.28},
    "Diversity of Installations/Platforms": {"Nominal": 1.00, ↙
        "High": 1.23, "Very_High": 1.52, "Extra_High": 1.87},
    "Recursive Levels in Design": {"Very_Low": 0.76, "Low": 0.87, ↙
        "Nominal": 1.00, "High": 1.21, "Very_High": 1.47},
    "Stakeholder Team Cohesion": {"Very_Low": 1.50, "Low": 1.22, ↙
        "Nominal": 1.00, "High": 0.81, "Very_High": 0.65},
    "Personnel/Team Capability": {"Very_Low": 1.50, "Low": 1.22, ↙
        "Nominal": 1.00, "High": 0.81, "Very_High": 0.65},
    "Personnel Experience/Continuity": {"Very_Low": 1.48, "Low": 1.22↙
        , "Nominal": 1.00, "High": 0.82, "Very_High": 0.67},
    "Process Capability": {"Very_Low": 1.47, "Low": 1.21, "Nominal": ↙
        1.00, "High": 0.88, "Very_High": 0.77, "Extra_High": 0.68},
    "Multisite Coordination": {"Very_Low": 1.39, "Low": 1.1 8, ↙
        "Nominal": 1.00, "High": 0.90, "Very_High": 0.80, ↙
        "Extra_High": 0.72},
    "Tool Support": {"Very_Low": 1.39, "Low": 1.1 8, "Nominal": 1.00↙
        , "High": 0.85, "Very_High": 0.72}
}


def phase_effort(effort):
    # Define the effort distribution percentages for each activity
    activity_percentages = {
        "Acquisition and Supply": .07,
        "Technical Management": .1 7,
        "System Design": .30,
        "Product Realization": .1 5,
        "Technical Evaluation": .31
    }

    # Calculate the sub-effort for each activity based on its ↙
        percentage
    sub_efforts = {activity: effort * percentage for activity, ↙
        percentage in activity_percentages.items()}

    return sub_efforts
```

```python
size_weights = {
    "System Requirements": {
        "Easy": 0.5,
        "Nominal": 1.00,
        "Difficult": 5.0
    },
    "Interfaces": {
        "Easy": 1.1 ,
        "Nominal": 2.8,
        "Difficult": 6.3
    },
    "Critical Algorithms": {
        "Easy": 2.2,
        "Nominal": 4.1 ,
        "Difficult": 11.5
    },
    "Operational Scenarios": {
        "Easy": 6.2,
        "Nominal": 14.4,
        "Difficult": 30
    }
}

def total_size(driver_counts):
    total = 0
    for driver, complexities in driver_counts.items():
        for complexity, count in complexities.items():
            total += count * size_weights[driver][complexity]
    return total
```

In summary, the entire chat, integration, and testing through the final iteration took 65 minutes. Doing it manually would have probably taken more than a day. All generated solutions were correct in this instance, and only minor refinements were made for clarity. However, generative AI solutions are non-deterministic even when using the same LLM. The caveat is always *Your Mileage May Vary*.

The cost model was subsequently extended into an online CGI application shown in Figure 5.3. In this case, generative AI was imperfect when converting it into a web-based application. It helped to create much of the underlying HTML but was not as effective integrating it with Python.

5.6 SOFTWARE ATTRIBUTES AND TRADE-OFFS

The software produced from a process has various attributes that affect its reliability and utility, which are crucial to consider. Attributes like execution speed, memory usage, portability, and maintainability are critical for producing software that can scale and adapt. Optimizing these attributes during the process is not straightforward, as choosing language features or libraries for one attribute often involves trade-offs

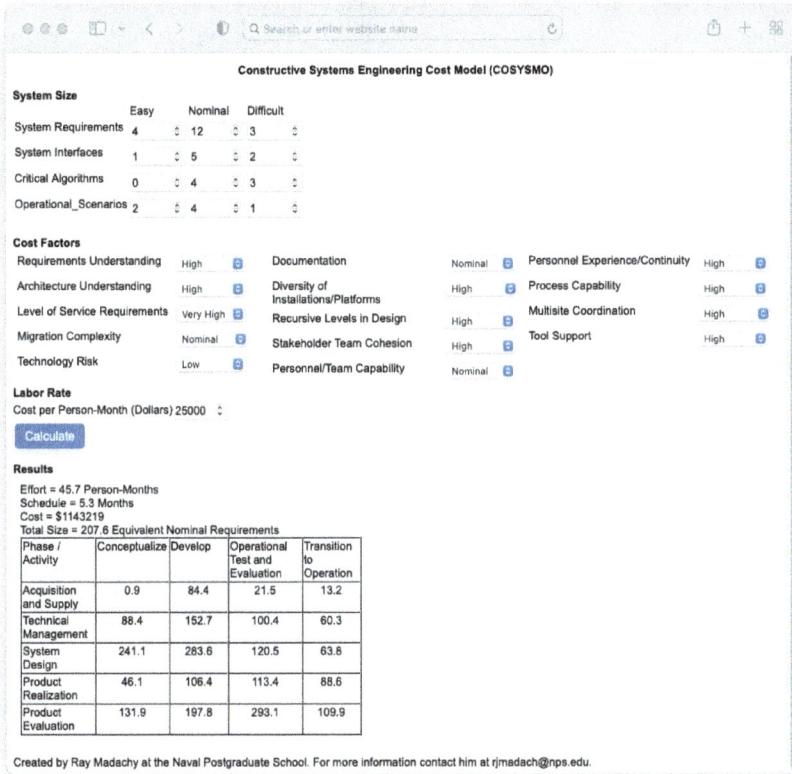

Figure 5.3 Online COSYSMO Tool Partially Developed with Generative AI

with others. For example, using NumPy can provide significant speed gains, but it may increase memory consumption and introduce platform dependencies. Conversely, using generators can reduce memory usage, but might slow execution due to the overhead of generating items on demand. Similarly, prioritizing maintainability may lead to more readable code, though potentially slower. Understanding these trade-offs allows for more informed decisions to strike the right balance between attributes based on the specific goals and requirements of the project.

Speed is often the most critical factor in many engineering applications. Table 5.3 provides a prioritized list of techniques that offer the highest potential for speed optimization (though not all techniques may be feasible in every scenario). Speed optimization focuses on reducing code execution time, especially for tasks involving large datasets or complex calculations. Techniques such as using NumPy for numerical computations or Just-in-Time (JIT) compilation with Numba can significantly speed up code by either converting Python into machine code or leveraging optimized C implementations. Additionally, built-in Python functions, list comprehensions, and asynchronous programming can all enhance performance.

The author conducted tests to measure the execution speed of different methods for generating values across various platforms. There can be significant variability both within sessions and between different IDE platforms. The representative results presented here compare the performance of NumPy versus Python lists for generating square numbers. To gather reliable statistics, a main loop in Listing 5.5 calls the NumPy and list-based functions for different numbers of square values.

Listing 5.5   Execution Speed Comparison of NumPy and Lists

```python
import numpy as np
import time

# List operation
def list_operation(n):
    return [i**2 for i in range(n)]

# NumPy operation
def numpy_operation(n):
    return np.arange(n)**2

sizes = [100, 1000, 10000, 100000]

multiplier = []
for n in sizes:
    list_times = []
    numpy_times = []
    for run in range(100):
        start = time.time()
        list_result = list_operation(n)
        list_times += [time.time() - start]
        start = time.time()
        numpy_result = numpy_operation(n)
        numpy_times.append(time.time() - start)
    multiplier += [np.mean(list_times) / np.mean(numpy_times)]
```

Figure 5.4 shows the distributions for 1000 runs to generate 1,000 and 10,000 values, respectively. In both cases, NumPy is significantly faster and exhibits less variability. As the number of generated values increases, the execution speed distribution for the list-based approach shifts higher by an order of magnitude, while the NumPy dispersion remains close to zero on the scale.

The performance difference becomes more pronounced with larger sequences. Figure 5.5 illustrates the impact of array size on performance, plotted on a logarithmic scale. The data points represent average speed improvement multipliers of NumPy over lists, calculated from 100 runs for each array size. The averages demonstrate improvement, but they still exhibit some variability. A more comprehensive analysis could involve adding an outer loop to collect multiple averages for each size, which could then be plotted as vertical boxplots to visualize the spread of values for each size.

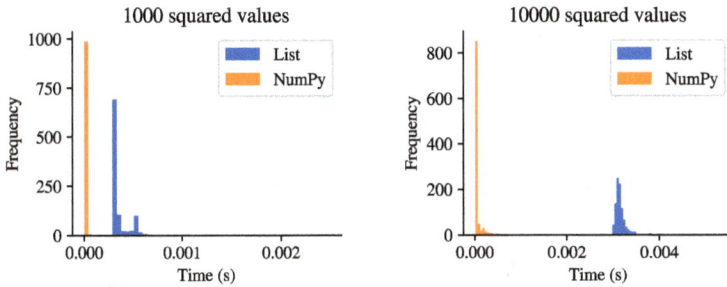

Figure 5.4   Execution Speed Histograms for NumPy and Lists

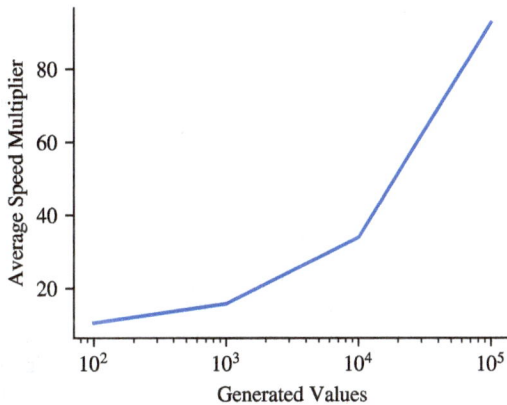

Figure 5.5   Execution Speed Multiplier for NumPy vs. Lists

Optimizing solely for speed often comes with trade-offs. For instance, parallel processing or JIT compilation can increase memory usage or limit portability. Sets and other optimized structures can consume more memory than lists due to the need for hashing and other optimizations, but they trade memory for performance.

Memory usage is sometimes a greater constraint than execution speed. Prioritized recommendations for reducing memory usage are provided in Table 5.4. Applications that handle large datasets or run on resource-constrained devices, such as embedded systems, will especially benefit from these optimizations. Reducing memory consumption can ensure that a program scales effectively. One of the most efficient techniques for memory optimization is using generators instead of lists, as they generate data on-the-fly without storing it all in memory. Similarly, in-place operations in

Table 5.3

Top Recommendations for Speed Optimization

Recommendation	Description
Use NumPy arrays for numerical computations	Significant speed improvement compared to lists due to vectorized operations implemented in C.
Use just-in-time compilation with Numba	Substantial speed boost for numerical code through just-in-time compilation, providing near C-like performance for array operations and loops.
Implement multiprocessing and multithreading for parallelism	Significant speed improvements for CPU-bound tasks by utilizing multiple cores.
Use built-in functions and libraries	Built-in functions offer faster performance, as they are optimized in C, making them much quicker than manual implementations for common tasks.
Utilize asynchronous programming with `asyncio`	Significant speed improvement for I/O-bound tasks by enabling non-blocking operations and concurrent task management. This is ideal for network and I/O-heavy applications.
Use list comprehensions and generator expressions	Faster than traditional `for` loops, especially for simple transformations or filtering tasks.
Optimize data structures	Choosing appropriate data structures improves efficiency. For example, dictionary lookups use a hash table and are faster than searching unsorted lists. Sets are the most efficient for membership checks, and tuples are faster when iterated over due to their immutability, which allows interpreter optimizations.
Minimize use of OOP for performance-critical code	Eliminates the overhead of method lookups and dynamic attribute access to improve execution speed.
Avoid unnecessary data copies in processing	When manipulating large datasets (e.g., NumPy arrays or Pandas dataframes), avoid creating unnecessary data copies to enhance performance. Use in-place operations on existing arrays or views where possible.
Use local variables over global variables	Local variables are quicker to access than global variables due to scope handling, resulting in improved performance, especially for frequently accessed variables within functions.

libraries like NumPy help avoid unnecessary memory copies, and more streamlined class designs can reduce the memory overhead of object-oriented code. However, some techniques, such as using generators, may lead to slower performance since they calculate values on demand rather than upfront.

Table 5.4

Top Recommendations for Memory Optimization

Recommendation	Description
Use generator expressions instead of lists for large data streams	Generators use no memory as they yield one item at a time, instead of holding all items in memory.
Use list comprehensions instead of loops	List comprehensions are more memory-efficient than manually appending to lists in a loop.
Optimize data structures	Choosing the appropriate data structure can improve memory efficiency. For example, using a set for membership testing avoids the overhead of storing duplicates.
Minimize use of OOP for performance-critical code	Reducing reliance on classes and objects can save memory by avoiding the overhead of storing instance attributes.
Use NumPy arrays instead of native Python lists for large datasets	NumPy arrays are more memory-efficient than Python lists for storing large datasets because they use contiguous memory and avoid Python's object overhead.
Use in-place operations	Perform in-place operations where possible (e.g., arr += 1 in NumPy) to reduce memory consumption by avoiding the creation of new arrays or dataframes.

Portable code operates seamlessly across different environments and platforms. Recommendations for achieving portability are provided in Table 5.5. Python's portability strengths come from its extensive standard library and cross-platform capabilities, though care must be taken when optimizing performance across systems. Built-in functions and libraries enhance portability since they are implemented in C and function consistently across platforms. Language features such as list comprehensions, generators, and asynchronous programming with asyncio are inherently portable across environments. However, using external libraries like NumPy or Numba for performance optimization may introduce platform-specific dependencies. Striking the right balance between platform-agnostic language features and performance-enhancing libraries is crucial.

Table 5.5

Top Recommendations for Portability

Recommendation	Description
Use built-in functions and libraries	Fully portable across all Python platforms and versions, with no external dependencies required.
Use list comprehensions and generator expressions	Core language features are fully portable across all Python platforms and versions.
Optimize data structures using Python's standard library	Data structures from Python's standard library are highly portable and work consistently across all Python versions and platforms.
Use `asyncio` for asynchronous programming	Fully portable for systems running Python 3.4 or later. Simplifies cross-platform asynchronous programming.
Ensure cross-platform compatibility for system-level tasks with `os` and `subprocess`	Python's `os` and `subprocess` modules ensure portability across different operating systems.
Write version-agnostic code	Use features compatible with multiple Python versions, especially for older systems, or utilize the `six` library to ease transitions between versions.
Avoid platform-specific libraries	Some libraries or tools may not be available across all platforms. Stick to widely supported libraries like NumPy, Pandas, and Python's standard library.
Use Python's virtual environments for dependency management	Ensures that projects are portable across different environments by isolating dependencies using `virtualenv` or Python's built-in `venv` module.

5.7   TEAM PROCESSES AND TOOLS

5.7.1   VERSION CONTROL AND COLLABORATION

Version control is a critical process in software engineering to manage changes to evolving software. It provides a systematic way to track modifications, revert to previous versions, collaborate on the same project without overwriting each other's work, and much more. Git is the most widely used version control system, while GitHub is a cloud-based platform for hosting and managing Git repositories. They have become standard in both open-source and commercial software development.

Utilizing version control is essential for managing projects of any size. It enables engineers to maintain a history of changes, collaborate effectively with team members, and ensure that every modification is accounted for. It helps maintain the integrity and consistency of software whether working on a personal project or contributing to a large-scale development.

GitHub extends the capabilities of Git by providing a platform for collaborative development. It allows multiple people to work on the same project by managing changes in a controlled manner. A key feature of GitHub is its support for branching and merging, which enables teams to work on different features or fixes in parallel without affecting the main codebase. Once the work on a branch is complete, it can be merged back into the main branch, with GitHub providing tools to resolve any conflicts that may arise.

5.7.2   CODE QUALITY AND STYLE

Maintaining consistent code quality and style across a project is essential for readability, maintainability, and collaboration. Code linting tools, such as `flake8` and `pylint`, analyze Python code for potential errors and enforce adherence to style guidelines. These tools help to identify common issues such as unused variables, improper indentation, and violations of naming conventions, which can lead to subtle bugs and make the code harder to read.

Automated formatting tools, such as `black`, can be used to standardize the appearance of code according to a consistent style guide. By applying formatting automatically, these tools remove the burden of manually maintaining code style. Incorporating linting and formatting ensures that all code commits meet quality standards before being merged into the main branch.

5.8   FUTURE OF SOFTWARE PROCESSES AND TOOLS

The landscape of Python, software processes, and tools will continue to evolve with advances in automation, AI, and new programming paradigms. We are in the early stages of generative AI for which the future can't be predicted. The tools and processes outlined in this chapter will remain relevant, but they will need to be adapted to meet the demands of increasingly complex and interconnected systems. The ongoing improvement of toolsets and the adoption of new technologies will be crucial for staying at the forefront.

As software development continues to become more collaborative and distributed, the importance of robust processes, such as version control, continuous integration, and automated testing, will increase. Engineers must be prepared to embrace these changes, continuously learning and adapting their practices to leverage the full potential of emerging tools and technologies. The future demands both technical proficiency and a deep understanding of the processes that drive successful software development. By staying informed and proactive, engineers can ensure that their work remains relevant and impactful in an ever-changing technological landscape.

5.9 SUMMARY

Command line tools and Python scripting can be used to enhance engineering productivity. Command line interfaces provide a direct way to interact with the operating system, run scripts, manage files, optimize code performance, manage dependencies, and control processes efficiently. Scripts can be written to automate repetitive tasks, manipulate data, and manage system resources. Magic commands in Jupyter Notebook environments also provide shortcuts to perform various operations. These capabilities are invaluable for both individual productivity and supporting team workflows.

Iterative development of software is done in small, manageable steps. By breaking a problem into pieces, individuals and teams can incrementally build, test, and refine their programs. This allows for continuous feedback and adaptation, which is particularly important on complex projects to reduce the risk of errors. Each iteration builds upon the previous ones, ensuring that the software evolves with increasing functionality and remains stable.

Generative AI can be leveraged to automate code generation, support debugging, and other activities to accelerate Python development. However, it's important to critically evaluate AI-generated solutions to ensure they are correct. Generated solutions may sometimes be overly complex, contain logical errors, or lack the nuance needed for specific applications. Testing becomes more crucial, and refining generated code requires an understanding of both Python and the application domain. Maintaining a balance between AI assistance and personal problem-solving skills will remain critical for successful software.

Testing and debugging are fundamental practices to ensure code reliability and correctness. Automated testing is important to create repeatable tests that can be run frequently to catch defects early. Unit tests should be designed to verify the functionality of individual components and be automated to save time and reduce testing errors. Debugging involves identifying and fixing defects, which can be supported with tools like pdb to step through code and examine the state of a program during execution. These practices help ensure that software behaves as expected and identifies issues to be addressed.

Version control systems like Git and platforms like GitHub are used for managing changes to artifacts over time. They enable teams to track modifications, revert to previous versions if necessary, and collaborate with others seamlessly. For team projects, version control is critical to help manage code changes from multiple contributors, prevent conflicts, and maintain a history of project evolution. GitHub also supports automating workflows for continuous integration, testing, and deployment.

Maintaining high code quality and consistent coding style fosters readability and maintainability. Code linting tools can identify potential errors and enforce coding standards. Code reviews can help find errors and ensure that code meets quality standards. These practices help prevent bugs, improve code readability, and facilitate easier maintenance and collaboration.

Proper documentation of design, functionality, and usage is crucial for understanding and maintaining software. Automating documentation generation can save time and ensure consistency, especially on large projects.

Optimizing performance involves understanding system trade-offs and making informed decisions to balance speed, memory usage, and maintainability. Attributes like execution speed, memory utilization, portability, and maintainability are crucial for working software that scales and can be adapted. Optimizing these attributes is not always straightforward as many optimizations require trade-offs. For example, code that is highly optimized for speed might consume more memory, or code designed to be highly portable may sacrifice some performance.

Profiling tools can help identify performance bottlenecks where code can be optimized. Handling large datasets and ensuring scalability are critical considerations in data-intensive applications. These practices ensure that software not only works correctly but also performs efficiently under various conditions.

Staying up-to-date with tools and best practices is essential to continuously improve and remain current. Toolsets should be regularly evolved to leverage new features and capabilities, enhance productivity and software quality. A mindset of staying on the vanguard and continuously refining processes is the best way to deal with upcoming engineering challenges. More complex aspects of the problems can then be focused on.

5.10   EXERCISES

1. Write a Python script to automate tasks in an engineering workflow, such as data collection, analysis, and archiving. Use Python or shell commands to copy or archive files from a specified directory. Next use Python to analyze these files, extract relevant data, and generate a summary report. After processing, move all analyzed files to an archive folder for storage.

2. Create a Python script that automates running tests on engineering analysis software undergoing changes. The script should use `!git` shell commands to pull the latest version of the repository, run Python unit tests (e.g., pytest), and trigger specific test cases with shell commands. Afterward, it should generate a summary report of the test results.

3. Develop a Python-based tool that automates the workflow for computational experiments. The script should perform multiple simulation runs with varying parameters by using shell commands to initiate the simulation software and modify input parameters. It should capture the output data, process it in Python, and generate comparative plots. All results should be archived in separate directories labeled with the experimental parameters used.

4. Write a Python script to automate data collection from a series of experiments or sensors, pre-processing, and machine learning process integration. The script should use shell commands to read from log files or sensors, and Python to clean the data, remove outliers, normalize it, and split it into training and testing sets. The script should then trigger a machine learning model training process, generate performance metrics, and save the results.

5. Write scripts to automate engineering team project workflows using GitHub's APIs. Automate tasks such as creating repositories, managing branches, and setting up pull requests. Implement GitHub Actions for continuous integration, ensuring that unit tests run automatically on each push or pull request.

6. Create a Python script to automate the creation of input files for external engineering analysis software (e.g., AutoCAD, ANSYS, Abaqus) or project management tools (e.g., Microsoft Project, JIRA). Use shell commands to organize files and Python to generate or modify input files based on user-defined parameters.

7. Develop a Python script to automatically post-process simulation results from Python or other tools. The script should use shell commands to extract relevant output files from the simulation software, and Python to parse the results and generate visualizations. It should automate the generation of comparative graphs for different simulation runs and export them as PDFs or images.

8. Create a control system for robotic maneuvers or drones, where AI assists in generating parts of the real-time control code. Start by coding basic kinematics or flight dynamics, and use AI to assist in generating code for specific movements or sensor integration. Test the AI-generated solutions in a simulation environment, and refine them for accuracy.

9. Take an existing Python script for engineering (from this book or elsewhere), and use generative AI to suggest improvements or refactor parts of the code for better performance. Test the AI-generated code for accuracy and analyze performance improvements.

5.11   ADVANCED EXERCISES

1. Use a quantitative weighted criteria approach (also called a Pugh matrix) to compare Python architecture and implementation options for a selected application. Choose and assign weights to relevant criteria such as speed, memory usage, portability, maintainability, and others. Score each option against each criterion on a selected scale (e.g., 0–10), and calculate the top choice based on the total weighted scores.

2. Participate in a Python open-source community, and contribute to a library relevant to engineering. You can propose improvements, perform development, engage in testing, or assist with other activities.

3. Select a complex engineering simulation and break it into small, manageable components. Use generative AI to assist with tasks like generating initial code, defining data structures, and plotting results. Evaluate the generated code by running unit tests and ensuring the outputs match expected values.

4. Take a large engineering project, and use AI to build automated testing suites. Prompt to specify different test cases, such as testing boundary conditions, performance under stress, or handling large datasets. Ensure the generated test cases accurately reflect the problem, and check for corner cases that could cause failures.

A Keywords and Built-in Functions

Python keywords are reserved words that form the core syntax and structure of the language shown in Table A.1. Python built-in functions available are listed alphabetically in Table A.2. Additional information on any function may be obtained from their docstring via the `function_name.__doc__` attribute.

Table A.1: Python Keywords

Keyword	Definition
and	Logical operator that returns True if both the operands are True
as	Create an alias name for a module, class, function, or variable
assert	Debugging aid that tests a condition, and triggers an error if the condition is not True
async	Modifier indicating a function is asynchronous, used with the await keyword
await	Used inside an asynchronous function to wait for a task to complete
break	Terminates the current loop and transfers execution to the statement immediately following the loop
class	Defines a new user-defined object type
continue	Causes the loop to skip the remainder of its body and immediately retest its condition prior to reiterating
def	Defines a function or method
del	Deletes an object, such as a variable, list item, or attribute of an object
elif	Conditional statement that follows an if statement and executes when the previous if statement condition is False
else	Conditional statement that follows an if statement and executes when the previous if statement condition is False
except	Block of code that is executed if an exception is raised in the try block
False	Special value representing the Boolean value False
finally	Block of code that is always executed after the try and except blocks, regardless of any exceptions raised
for	Loop that is executed a specific number of times, iterating over a sequence of values or elements in a collection

Continued on next page

Table A.1: Python Keywords – Continued

Keyword	Definition
from	Imports a specific module or object from a module
global	Declaration indicating a variable is global, i.e. accessible from anywhere in the code
if	Conditional statement that tests a condition and executes a block of code if the condition is True
import	Imports a module into the current namespace
in	Tests membership in a collection, such as a list or dictionary
is	Comparison operator that tests object identity
lambda	Small anonymous function that can take any number of arguments but can only have one expression
None	Special value representing the absence of a value or a null value
nonlocal	Declaration indicating a variable is nonlocal; i.e. it is not local to the current function
not	Logical operator that negates the boolean value of an expression
or	Logical operator that returns True if either of the operands are True
pass	Null statement that does nothing and is used as a placeholder where a statement is required by the syntax
raise	Raises an exception or error
return	Returns a value from a function or method
True	Special value representing the Boolean value True
try	Block of code that is executed and monitored for exceptions
while	Loop that repeatedly executes a block of code as long as the condition is True
with	A context manager that automatically takes care of acquiring and releasing resources, such as file handles
yield	Used in generator functions to return a value and pause execution, to be resumed later

Table A.2: Built-in Functions

Function	Description
abs()	Returns the absolute value of a number
all()	Returns True if all items in an iterable object are true
any()	Returns True if any item in an iterable object is true
ascii()	Returns a readable version of an object. Replaces non-ascii characters with escape characters
bin()	Returns the binary version of a number
bool()	Returns the Boolean value of the specified object
bytearray()	Returns an array of bytes
bytes()	Returns a bytes object
callable()	Returns True if the specified object is callable, otherwise False
chr()	Returns a character from the specified Unicode code.
classmethod()	Converts a method into a class method
compile()	Returns the specified source as an object, ready to be executed
complex()	Returns a complex number
delattr()	Deletes the specified attribute (property or method) from the specified object
dict()	Returns a dictionary (Array)
dir()	Returns a list of the specified object's properties and methods
divmod()	Returns the quotient and the remainder when argument1 is divided by argument2
enumerate()	Takes a collection (e.g. a tuple) and returns it as an enumerate object
eval()	Evaluates and executes an expression
exec()	Executes the specified code (or object)
filter()	Uses a filter function to exclude items in an iterable object
float()	Returns a floating point number
format()	Formats a specified value
frozenset()	Returns a frozenset object
getattr()	Returns the value of the specified attribute (property or method)
globals()	Returns the current global symbol table as a dictionary
hasattr()	Returns True if the specified object has the specified attribute (property/method)
hash()	Returns the hash value of a specified object
help()	Executes the built-in help system
hex()	Converts a number into a hexadecimal value
id()	Returns the id of an object
input()	Allows user input

Continued on next page

Table A.2: Built-in Functions – Continued

Function	Description
int()	Returns an integer number
isinstance()	Returns True if a specified object is an instance of a specified object
issubclass()	Returns True if a specified class is a subclass of a specified object
iter()	Returns an iterator object
len()	Returns the length of an object
list()	Returns a list
locals()	Returns an updated dictionary of the current local symbol table
map()	Returns the specified iterator with the specified function applied to each item
max()	Returns the largest item in an iterable
memoryview()	Returns a memory view object
min()	Returns the smallest item in an iterable
next()	Returns the next item in an iterable
object()	Returns a new object
oct()	Converts a number into an octal
open()	Opens a file and returns a file object
ord()	Converts an integer representing the Unicode of the specified character
pow()	Returns the value of x to the power of y
print()	Prints to the standard output device
property()	Gets, sets, deletes a property
range()	Returns a sequence of numbers, starting from 0 and increments by 1 (by default)
repr()	Returns a readable version of an object
reversed()	Returns a reversed iterator
round()	Rounds a numbers
set()	Returns a new set object
setattr()	Sets an attribute (property/method) of an object
slice()	Returns a slice object
sorted()	Returns a sorted list
staticmethod()	Converts a method into a static method
str()	Returns a string object
sum()	Sums the items of an iterator
super()	Returns an object that represents the parent class
tuple()	Returns a tuple
type()	Returns the type of an object
vars()	Returns the __dict__ property of an object
zip()	Returns an iterator, from two or more iterators

References

1. ArduPilot Dev Team. Ardupilot documentation, 2024. URL: https://ardupilot.org/ardupilot/index.html.

2. Stephen Cass. Top Programming Languages 2022. *IEEE Spectrum*, 2022. URL: https://spectrum.ieee.org/top-programming-languages-2022.

3. Allen B Downey. *Think Python: How to Think Like a Computer Scientist*. Green Tea Press, 2nd edition, 2015.

4. Python Software Foundation. Welcome to python.org, 2024. URL: http://www.python.org/.

5. John Guttag. *Introduction to Computation and Programming Using Python: With Application to Understanding Data*. MIT Press, 2nd edition, 2016.

6. Robert Johansson. *Numerical Python: Scientific Computing and Data Science Applications with NumPy, SciPy and Matplotlib*. Apress, Division of Springer Nature, New York, NY, 2019.

7. Jaan Kiusalaas. *Numerical Methods in Engineering with Python 3*. Cambridge University Press, New York, NY, 2013.

8. Qingkai Kong, Timmy Siauw, and Alexandre Bayen. *Python Programming and Numerical Methods: A Guide for Engineers and Scientists*. Academic Press, Elsevier, Cambridge , MA, 2021.

9. Hans P. Langtangen. *A Primer on Scientific Programming with Python*. Springer, Berlin, 5th edition, 2016.

10. Raymond J. Madachy. *Software Process Dynamics*. Wiley-IEEE Computer Society Press, Hoboken, NJ, 2007.

11. Raymond J. Madachy and Daniel X. Houston. *What Every Engineer Should Know About Modeling and Simulation*. CRC Press, Boca Raton, FL, 2017.

12. Matplotlib development team. Examples - matplotlib 3.8.2 documentation, 2024. URL: https://matplotlib.org/stable/gallery/index.html.

13. Wes McKinney. *Python for Data Analysis: Data Wrangling with Pandas, NumPy, and IPython*. O'Reilly Media, Sebastopol, CA, 2nd edition, 2017.

14. NumPy Developers. NumPy Documentation – NumPy v2.0 Manual, 2024. URL: https://numpy.org/doc/stable/.

15. Travis E. Oliphant. Python for scientific computing. *Computing in Science & Engineering*, 9(3):10–20, 2007.

16. PyTorch Contributors. Pytorch, 2021. Version 1.8. URL: https://pytorch.org/.

17. scikit-learn Developers.   scikit-learn, 2024.   Version 0.24.   URL: https://
    scikit-learn.org/.

18. SciPy Developers. SciPy, 2024. URL: https://www.scipy.org/.

19. se-lib Development Team. Examples - Simulation, 2024. URL: http://se-lib.org/
    examples.html#simulations.

20. statsmodels   Developers.   statsmodels.regression.linear_model.ols,   2024.
    URL: https://www.statsmodels.org/stable/generated/statsmodels.
    regression.linear_model.OLS.html.

21. Al Sweigart. *Automate the Boring Stuff with Python: Practical Programming for Total
    Beginners*. No Starch Press, San Francisco, CA, 1st edition, 2015.

Index

For Product Safety Concerns and Information please contact our EU
representative GPSR@taylorandfrancis.com
Taylor & Francis Verlag GmbH, Kaufingerstraße 24, 80331 München, Germany

www.ingramcontent.com/pod-product-compliance
Lightning Source LLC
Chambersburg PA
CBHW060331220326
41598CB00023B/2675